A LITTLE HISTORY OF SCIENCE

For Alex and Peter

WILLIAM BYNUM

A LITTLE HISTORY OF SCIENCE

YALE UNIVERSITY PRESS
NEW HAVEN AND LONDON

For information about this and other Yale University Press publications, please contact:
U.S. Office: sales.press@yale.edu   www.yalebooks.com
Europe Office: sales@yaleup.co.uk   www.yalebooks.co.uk

Set in Minion Pro by IDSUK (DataConnection) Ltd
Printed in the United States of America

Library of Congress Cataloging-in-Publication Data

Bynum, W. F. (William F.), 1943–
  A little history of science / William Bynum.
    pages cm
    Includes bibliographical references and index.
    ISBN 978–0–300–13659–3 (hardback)
  1. Science—History. 2. Science and civilization. I. Title.
  Q125.B98 2012
  509—dc23

                                                        2012026738

A catalogue record for this book is available from the British Library.

10 9 8 7 6 5 4 3 2 1

Contents

CHAPTER 1

In the Beginning

Science is special. It's the best way we have of finding out about the world and everything in it – and that includes us.

People have been asking questions about what they have seen around them for thousands of years. The answers they have come up with have changed a lot. So has science itself. Science is dynamic, building upon the ideas and discoveries which one generation passes on to the next, as well as making huge leaps forward when completely new discoveries are made. What hasn't changed is the curiosity, imagination and intelligence of those doing science. We might know more today, but people who thought deeply about their world 3,000 years ago were just as smart as we are.

This book isn't only about microscopes and test tubes in laboratories, although that is what most people think of when they think of science. For most of human history, science has been used alongside magic, religion and technology to try to understand and control the world. Science might be something as simple

as observing the sun rise each morning, or as complicated as identifying a new chemical element. Magic could be looking at the stars
to foretell the future, or maybe what we would call a superstition,
like keeping out of the path of a black cat. Religion might lead you
to sacrifice an animal to appease the gods, or to pray for world
peace. Technology might involve knowing how to light a fire or
build a new computer.

Science, magic, religion and technology were used by the earliest
human societies that settled in river valleys across India, China
and the Middle East. The river valleys were fertile, which allowed
crops to be planted each year, enough to feed a large community.
This allowed some people in these communities enough time to
focus on one thing, to practise and practise, and become expert at
it. The first 'scientists' (though they wouldn't have been called that
at the time) were probably priests.

In the beginning, technology (which is about 'doing') was more
important than science (which is about 'knowing'). You need to
know what to do, and how to do it, before you can successfully grow
your crops, make your clothes, or cook your food. You don't need
to know *why* some berries are poisonous, or some plants edible, to
learn how to avoid the one and grow the other. You don't have to
have a reason why the sun rises each morning and sets each evening,
for these things to happen, each and every day. But human beings
are not only able to learn things about the world around them, they
are also curious, and that curiosity lies at the heart of science.

We know more about the people of Babylon (in present-day
Iraq) than we do about other ancient civilisations, for a simple
reason: they wrote on clay tablets. Thousands of these tablets,
written almost 6,000 years ago, have survived. They tell us how the
Babylonians viewed their world. They were extremely organised,
keeping careful records of their harvests, stores, and state finances.
The priests spent much of their time looking after the facts and
figures of ancient life. They were also the main 'scientists', surveying
land, measuring distances, viewing the sky, and developing techniques for counting. We still use some of their discoveries today.
Like us, they used tally marks to keep count; this is when you make

four vertical marks and cross through these diagonally with a fifth, which you might have seen in cartoons of a prison cell, made by the prisoners keeping count of how many years they have been locked up. Far more importantly, it was the Babylonians who said there should be sixty seconds in a minute and sixty minutes in an hour, as well as 360 degrees in a circle and seven days in the week. It is funny to think that there is no real reason why sixty seconds make a minute, and seven days make a week. Other numbers would have worked just as well. But the Babylonian system got picked up elsewhere and it has stuck.

The Babylonians were good at astronomy – that is, examining the heavens. Over many years they began to recognise patterns in the positions of stars and planets in the sky at night. They believed that the earth was at the centre of things, and that there were powerful – magical – connections between us and the stars. As long as people believed that the earth was the centre of the universe, they didn't count it as a planet. They divided the night sky into twelve parts, and gave each part a name associated with certain groups (or 'constellations') of stars. Through a heavenly game of Join-the-dots, the Babylonians saw pictures of objects and animals in some constellations, such as a set of scales and a scorpion. This was the first Zodiac, the basis of astrology, which is the study of the influence of the stars upon us. Astrology and astronomy were closely linked in ancient Babylon and for many centuries after-wards. Many people today know which sign of the Zodiac they were born under (I am a Taurus, the bull) and read their horo-scopes in newspapers and magazines for advice about their lives. But astrology is not part of modern science.

The Babylonians were only one of several powerful groups in the ancient Middle East. We know most about the Egyptians, who settled along the River Nile as early as 3,500 BC. No civilisation before or since was so dependent on a single natural feature. The Egyptians relied on the Nile for their very existence, for every year as the mighty river flooded it brought rich silt to replenish the land around its banks, and so prepare it for the next year's crops. Egypt is very hot and dry, so a lot of things have survived for us to admire and learn from today, including many pictures, and a kind of pictorial writing, called hieroglyphics. After Egypt was conquered first by the Greeks and then by the Romans, the ability to read and write hieroglyphs disappeared, and so for almost 2,000 years the meaning of their writing was lost. Then, in 1798, a French soldier found a round tablet in a pile of old rubble in a little town near Rosetta, in the north of Egypt. It had a proclamation written in three languages: hieroglyphics, Greek, and an even older form of Egyptian writing called demotics. The Rosetta Stone came to London, where you can see it today in the British Museum. What a breakthrough! Scholars could read the Greek and therefore translate the hieroglyphs, decoding the mysterious Egyptian writing. Now we could really begin to learn about the ancient Egyptians' beliefs and practices.

Egyptian astronomy was similar to the Babylonians', but Egyptian concern with the afterlife meant that they were more practical in their stargazing. The calendar was very important, not only to tell them when it was the best time to plant, or when to expect the Nile to flood, but also to plan religious festivals. Their 'natural' year was 360 days – that is, twelve months made up of three weeks lasting ten days each – and they added an extra five days at the end of the year to keep the seasons from slipping. The Egyptians thought that the universe was shaped like a rectangular box, with their world at the base of the box, and the Nile flowing exactly through the centre of that world. The beginning of their year coincided with the flooding of the Nile, and they eventually linked it with the nightly rising of the brightest star in the night sky, which we call Sirius.

As in Babylon, the priests were important in the courts of the Pharaohs, the Egyptian rulers. The Pharaohs were considered to be divine, and able to enjoy a life after death. This is one reason why they constructed the pyramids, which are really gigantic funeral monuments. Pharaohs, their relatives and other important people, along with servants, dogs, cats, furniture and food supplies, were placed in these massive structures to await new life in the next world. To preserve the bodies of important people (after all, it wouldn't do to turn up in the afterlife rotten and stinking) the Egyptians developed ways of embalming the dead. This meant first of all removing the internal organs (they had a long hook for scooping out the brain through the nostrils) and placing them in special jars. Chemicals were used to preserve the rest of the body, which was then wrapped in linen and put in its final resting place in its tomb.

Embalmers would have had a pretty good idea of what the heart, lungs, liver and kidneys looked like. Unfortunately, they did not describe the organs that they removed, so we don't know what they thought the organs did. However, other medical papyri have survived, and these do tell us about Egyptian medicine and surgery. As was common at the time, the Egyptians believed a mix of religious, magical and natural things caused diseases. Healers would have recited spells while giving their remedies to patients. But many of the cures invented by the Egyptians do seem to have come from careful observation of illnesses. Some of the medicines they put on dressings for wounds after injury or surgery might well have kept the wound free from germs and so aided healing. This was thousands of years before we even knew what germs were.

At this stage of history, counting, astronomy and medicine were the three most obvious 'scientific' fields of activity. Counting, because you need to know 'how many' before you can plant enough crops and trade with other people, or to see if you have enough soldiers or pyramid builders at your disposal. Astronomy, because the sun, moon and stars are so closely related to the days, months and seasons, that carefully noting their positions is fundamental for calendars. Medicine, because when people fall sick or are injured, they naturally seek help. But in each of these cases, magic, religion, technology and science were mixed, and for these ancient Middle Eastern civilisations, we have to guess a lot about why people did what they did, or how ordinary people went about their daily lives. Ordinary people are always difficult to know about, since it was mostly powerful people, who could read and write, who have left the records of history. This was also true in two other ancient civilisations that started at about the same time, but in faraway Asia: China and India.

CHAPTER 2

Needles and Numbers

Keep on travelling eastwards from Babylon and Egypt and you'll find lands where ancient civilisations flourished on either side of the rocky Himalayas, in India and China. Some 5,000 years ago people were living there in towns and cities ranged along the Indus and Yellow River valleys. In those days, India and China were both immense territories, even larger than they are today. Both were part of vast overland and overseas trading networks – channelled along the spice routes – and their people had developed writing and science to a high level. The one helped the other: science benefited trade, and the wealth from trade allowed the luxury of study. In fact until about 1500, science in each of these civilisations was at least as advanced as in Europe. India gave us our numbers and a love of mathematics. From China came paper and gunpowder and that indispensible gadget for navigation: the compass.

Today, China is a major force in the world. Things like clothes, toys and electronic goods made there are sold all over the globe:

check the label in your trainers. For centuries, however, people in the West looked at this vast country with amusement or suspicion. The Chinese did things their way; their country seemed both mysterious and unchanging.

We now know that China was always a dynamic country, and that its science too was constantly changing. But one thing remained unchanged there over the centuries: writing. Chinese writing is made up of ideographs, little pictures that represent objects, which look strange to alphabet users like us. But if you know how to interpret the little pictures, it means you can read old – very old – Chinese texts as easily as you can read more recent writings. In fact, we have China to thank for the invention of paper, which made writing much easier. The oldest example we know about dates from about AD 150.

Ruling China was never easy, but science could help. Perhaps the greatest-ever engineering project, the Great Wall of China, was begun during the fifth century BC, during the Eastern Zhou Dynasty. (Chinese history is divided into dynasties, associated with powerful rulers and their courts.) The Wall was meant to keep the barbarians from the north out, as well as to keep the Chinese in! It took centuries to complete, being constantly extended and repaired, and now runs for 5,500 miles. (For some years people thought the Wall could be seen from space, but it's not true: China's own astronaut failed to spot the structure.) Another remarkable engineering feat, the Grand Canal, was started under the Sui Dynasty, in the fifth century. Making use of some natural waterways en route, the thousand-mile Canal connected the large inland city of Beijing in the north with Hangzhou on the southern coast, and from there to the outside world. These monuments are powerful reminders of the skills of Chinese surveyors and engineers, but also of the tremendous amount of hard human labour their construction needed. The Chinese had invented the wheelbarrow, but labourers still had to dig, push and build.

The Chinese viewed the universe as a kind of living organism, in which forces connected everything. The fundamental force, or energy, was called Qi (pronounced Chee). Two other basic forces were *yin* and *yang*: yin, the female principle, was associated with darkness, clouds and moisture; yang, the male principle, with ideas of sunshine, heat and warmth. Things are never either all yin or all yang – the two forces are always combined in various degrees. According to Chinese philosophy, each of us has some yin and some yang, and the exact combination affects who we are and how we behave.

The Chinese believed that the universe was made up of five elements: water, metal, wood, fire and earth. These elements were not simply the ordinary water or fire that we see around us, but principles that went together to compose the world and the heavens. Each had different characteristics, of course, but also interlocking powers, much like Transformer toys. For example, wood could overcome earth (a wooden spade can dig it); metal could chisel wood; fire could melt metal; water could extinguish fire; and earth could dam water. (Think of the game Paper, Scissors, Stone, actually invented in China.) These elements, combined with the forces of yin and yang, produce the cyclical rhythms of time and nature, the seasons, cycles of birth and death, and the movements of the sun, stars and planets.

Since everything is made up of these elements and forces, everything is in some sense alive and joined. So a notion of the 'atom' as a basic unit of matter never developed in China. Nor did natural philosophers there think that they had to express everything with numbers in order for it to be 'scientific'. Arithmetic was very practical: doing your sums when you were buying and selling, weighing goods, and so on. The abacus, a device with sliding beads on wires that you might have learned to count on, was written about in the late 1500s. It was probably invented earlier. An abacus speeds up counting, as well as addition, subtraction, multiplication and division.

Numbers were also used to calculate the length of the days and years. As early as 1400 BC, the Chinese knew that the year is 365¼ days long, and, like most early civilisations, they used the moon to calculate the months. As with all ancient peoples, the Chinese measured a year as being the length of time it took for the sun to return to its starting point in the sky. The cycles in the movements of planets like Jupiter, and of the stars, fit nicely into the idea that everything in nature is cyclical. The 'Supreme Ultimate Grand Origin' was an immense calculation to find out how long it would take for the whole universe to make a complete cycle: 23,639,040 years. This meant that the universe was very old (though we now know it is much older). The Chinese also thought about how the universe was structured. Some of the early Chinese star maps show that they understood how to represent, on a two-dimensional map, things that exist in a curved space. Xuan Le, who lived in the later Han Dynasty (AD 25–220), believed that the sun, moon and stars floated in empty space, driven by the winds. This was very different to the ancient Greek belief that these heavenly bodies were fixed in solid spheres, and is much more like how we understand space today. Stargazers in China recorded unusual events very carefully, so their records, since they go back so far, are still useful to modern astronomers.

Since the Chinese believed that the earth was very old, they had no difficulty recognising fossils as the hardened remains of plants and animals that had once been alive. Stones were grouped according to such things as hardness and colour. Jade was especially prized, and craftsmen carved pieces of jade into beautiful statues. Earthquakes are common in China, and although no one could explain why they occur, in the second century AD, a very learned man called Zhang Heng used a hanging weight that swayed when the earth shook, to record the earth's tremors. This was a very early version of what we call a seismograph, a machine which draws a straight line until the earth moves, when it wiggles.

Magnetism was understood for practical purposes. The Chinese learned how to magnetise iron by heating it to a high temperature and letting it cool while it was pointing in a north–south direction. China had compasses long before they were known in the West, and they were used both for navigation and for fortune telling. Most commonly, they were 'wet': just a magnetised needle floating in a bowl of water. We are used to saying that compass needles point north, but for the Chinese they pointed south. (Of course our compasses point south too – just with the opposite end of the needle. It doesn't really matter which direction you choose, as long as everyone agrees on it.)

The Chinese were skilled chemists. Many of the best were Taoists, members of a religious group who followed Lau Tsu, who lived some time between the sixth and the fourth century BC. (*Tao* means 'Way' or 'Path'.) Others followed Confucius or the Buddha. The philosophies of these religious leaders influenced the attitudes of their followers towards the study of the universe. Religion has always influenced how people view their surroundings.

The chemistry that the Chinese were able to perform was quite sophisticated for their time. For instance, they could distil alcohol and other substances, and could extract copper from solutions. By blending charcoal, sulphur and potassium nitrate, they made gunpowder. This was the first chemical explosive and the springboard for the invention of both fireworks and weapons. You could say gunpowder showed the yin and yang of the chemical world: it exploded prettily in tremendous firework displays at court while also being used to fire guns and cannons on eastern battlefields as early as the tenth century. It is not certain exactly how the recipe and instructions for making this powerful substance arrived in Europe, but there is a description of it there in the 1280s. It gradually made war everywhere more deadly.

The Chinese also had alchemists, who sought the 'elixir of life': some substance that would increase how long we can live, or maybe even make us immortal. (There is more about alchemy in Chapter 9.) They failed to find it, and in fact several emperors would have lived longer had they *not* taken these experimental, poisonous 'cures'. However, searching for this magical substance revealed many drugs that could be used to treat ordinary diseases. As in Europe, Chinese doctors used extracts of plants to treat diseases, but they also made compounds from sulphur, mercury and other substances. The Artemisia plant was used to treat fevers. It was made into an extract and burnt on the skin at specific points to aid the flow of the 'vital juices'. The recipe and method were recently discovered in a book about drugs written around 1,800 years ago. Tested in a modern laboratory, it was found to be effective against malaria, a leading cause of death today in tropical countries. One of the symptoms of malaria is a high fever.

Medical books began to be written in China as early as the second century BC, and ancient Chinese medicine lives on throughout the world today. Acupuncture, which involves sticking needles into certain areas of the skin, is widely practised to help cure disease, deal with stress, and ease pain. It is based on the idea that the body has a series of channels through which the Qi energy flows, and so the acupuncturist uses the needles to stimulate or unblock these channels. Sometimes operations are carried out with little more than the needles inserted into the patient's body to block out the pain. Modern Chinese scientists work just like their colleagues in the West, but Traditional Chinese Medicine (TCM) still has many followers all over the world.

So does traditional Indian medicine. It is called *Ayurveda*, and is based on works known by this name written in the ancient language, Sanskrit, between about 200 BC and AD 600. Ayurveda taught that there are fluids in the body, called *doshas*. There were three of these: *vata* is dry, cold and light; *pitta* is hot, sour and pungent; and *kapha* is cold, heavy and sweet. These doshas are necessary for the proper functioning of our bodies, and when there is too much or too little of one or more of them, or when they are in the wrong place, disease occurs. Inspecting the patient's skin, and feeling the pulse, were also very important as the Indian doctor tried to decide what the disease was. Drugs, massage and special diets could correct the imbalance. Indian doctors used the juice from the poppy, which produces the drug opium, to calm their patients and relieve their pain.

One other ancient Indian medical work, the *Susruta*, concentrated on surgery. Some of the operations it describes are remarkably delicate for this early period. For example, when a patient suffered from a cataract (a clouding in the lens of the eye which makes it hard to see) the surgeon would gently stick a needle into the eyeball and push the cataract to one side. Indian surgeons also

used flaps of the patient's own skin to repair damaged noses, probably the earliest instance of what we call plastic surgery.

This Ayurvedic medicine was associated with Hindu practitioners. When Muslims also settled in India in about 1590, they
brought their own medical ideas, based on ancient Greek medicine
interpreted by early Islamic doctors. This medicine, called *Yunani*
(which means 'Greek'), developed side by side with the Ayurvedic
system. Both continue to be used in India today alongside the
medicine we are all familiar with – Western medicine.

India had its own scientific traditions. Stargazers in India made
sense of the heavens, the stars, sun and moon by drawing on the
work of the Greek astronomer Ptolemy, and some scientific works
from China that had been brought back by Indian Buddhist
missionaries. There was an observatory at Ujjain, and one of
the earliest Indian scientists whose name we know, Varahamihira
(*c.* 505), worked there. He collected old astronomical works and
added his own observations. Much later, in the sixteenth century,
observatories were built at Delhi and Jaipur. The Indian calendar
was quite accurate, and like the Chinese, Indians believed that the
earth was very old. One of their astronomical cycles was 4,320,000
years long. Indians shared in the search for an elixir that would
convey long life. They also looked for a way to create gold from
ordinary metals. But the most important contribution made by
Indian science was in mathematics.

It is from India, via the Middle East, that we have the numbers
we call 'Arabic': the familiar 1, 2, 3 and so forth. The idea of 'zero'
first came from India, too. Along with the numbers we still use,
Indian mathematicians also had the basic idea of 'place-holding'.
Take a number like 170. The '1' = 100, it holds the 'hundreds' place;
the '7' = 70, it holds the 'tens' place; and the zero holds the 'units'
place. It comes so naturally to us that we don't even think about it,
but if we didn't have place-holding, writing large numbers would
be much more complicated. The most famous ancient Indian
mathematician, Brahmagutpa, who lived in the seventh century,
worked out how to calculate the volumes of prisms and other
figures. He was the first person to mention the number '0', and

knew that anything multiplied by 0 is 0. It took almost 500 years before another Indian mathematician, Bhaskara (b. 1115) pointed out that anything *divided* by 0 would be infinity. Modern mathematical explanations of the world would be impossible without these concepts.

Whereas traditional medical systems in India and China still compete with Western medicine, in science it's different. Indian and Chinese scientists work with the same ideas, tools and aims as their colleagues elsewhere in the world. Whether in Asia or anywhere else, science now *is* a universal science, which developed in the West.

But remember that we got numbers from India, and paper from China. Write out the '9 times' table, and you are using gifts that are very old, and from the East.

CHAPTER 3

Atoms and the Void

In about 454 BC, the Greek historian Herodotus (*c.* 485–425 BC) visited Egypt. Just like us, he was astonished by the Pyramids, and by the gigantic statues – sixty feet tall – at Thebes further up the Nile. He could not quite believe just how old everything was. Egypt's glory had passed and it had already long ago been overrun by the Persians. Herodotus was living in a much younger, more vigorous society that was still on the up, and one that would conquer Egypt a century later under Alexander the Great (356–323 BC).

By Herodotus's time, people who thought and wrote in Greek dominated a growing part of the eastern Mediterranean. They had written down the works of Homer, the blind poet, such as the story of how the Greeks defeated the Trojans by building and hiding in a giant horse, as well as the fantastic journey home of the Greek soldier Odysseus, who had masterminded the Trojan War. The Greeks were great ship-builders, traders and thinkers.

One of the earliest of these thinkers was Thales (*c.* 625–545 BC), a merchant, astronomer and mathematician of Miletus, on the coast of what is now Turkey. Nothing that he wrote survives directly, but later authors quote him as well as telling anecdotes that illustrate what he was like. One of them says that he was once so busy looking up at the stars that he forgot to look where he was stepping and fell down a well. In another story, Thales comes out on top: because he was clever, he was able to see that there was going to be a very big olive harvest. He hired all the olive presses long before the harvest, when no one needed them, and then when the harvest came in, he was able to rent them out for a large profit. Thales was not the first absent-minded professor – and we shall meet more later – nor the only one to make money from applying his science.

Thales was said to have visited Egypt and brought back Egyptian mathematics to the Greeks. This may be just another story, like the one about him correctly predicting a total eclipse of the sun (he didn't know enough astronomy to do that). More likely, though, was the way he tried to explain many natural occurrences, such as the fertilisation of the land by the flooding of the Nile, and the way earthquakes are caused by the overheating of water inside the earth's crust. For Thales, water was the chief element, and he pictured the earth as a disc floating on an enormous ocean. That sounds very funny to us, but the point is that Thales really wanted to explain things in natural, rather than supernatural, terms. The Egyptians thought that the Nile flooded because of the gods.

Unlike Thales, Anaximander (*c.* 611–547 BC), also from Miletus, believed that fire was the most important substance in the universe. Empedocles (*c.* 500–430 BC), from Sicily, came up with the idea of there being four elements: air, earth, fire and water. That idea is familiar to us because it became the default mode of thinkers for almost 2,000 years, until the end of the Middle Ages.

Being the default mode doesn't mean absolutely everyone accepted the four-element scheme as the last word. Also in Greece, and later in Rome, a group of philosophers known as the atomists believed that the world is actually made up of tiny particles called atoms. The most famous of these early atomists was Democritus, who lived around 420 BC. What we know of his ideas comes from a few fragments of his thought which other authors quoted. Democritus thought that in the universe there were lots of atoms, and that they had always existed. Atoms could not be broken down any further, nor could they be destroyed. Although they were far too small to be seen, he believed that they must be of different shapes and sizes, for this would explain why the things made of atoms have different tastes, textures and colours. But these larger things only exist because we humans taste, feel and see. In reality, Democritus insisted, there is nothing but 'atoms and the void', what we call matter and space.

Atomism was not all that popular, especially Democritus's and his followers' view of how living beings 'evolved' through a kind of trial and error. One funny version suggested that there had once been a large number of the various parts of plants and animals that could potentially join up in any kind of combination – an elephant's trunk could attach to a fish, a rose petal to a potato, and so forth – before they finally all fitted together in ways that we see now. The idea was that if a dog's leg accidentally joined up to a cat, that animal would not survive and so there would be no cats with dogs' legs. After a period of time, therefore, all the dogs' legs ended up on dogs, and – thankfully – all the human legs ended up on humans. (Another ancient Greek version of evolution seems more realistic, if still a little icky: all living things were supposed to have gradually come into existence from a very ancient slime.)

Because atomism doesn't see any final purpose or great design in the universe, with things just happening by luck and necessity, most people didn't like it. It is a pretty bleak view, and most Greek philosophers sought purpose, truth and beauty. The Greeks who lived at the same time as Democritus and his fellow atomists would have heard their full arguments; what we know of them is only through quotations and discussions of philosophers who came later. One atomist who lived in Roman times, Lucretius (c. 100–c. 55 BC), wrote a beautiful scientific poem, *De rerum natura* ('On the nature of things'). In this poem he described the heavens, the earth and everything on the earth, including the evolution of human societies, in terms of atomism.

We know the names and some of the contributions of dozens of ancient Greek scientists and mathematicians over a period of almost a thousand years. Aristotle was one of the greatest. His view of nature was so powerful that it dominated long after his death (and we'll turn to him in Chapter 5). But three people who lived after Aristotle made especially significant contributions to the ongoing development of science.

Euclid (c. 330–c. 260 BC) was not the first person to think about geometry (the Babylonians were pretty good at it). But he was the one who brought together, in a kind of textbook, the basic assumptions, rules and procedures of the subject. Geometry is a very practical kind of mathematics which deals with space: points, lines, surfaces, volumes. Euclid described geometrical ideas such as the way parallel lines never meet, and how the angles of a triangle add up to 180 degrees. His great book, *Elements of Geometry*, was admired and studied across Europe. You might study his 'plain geometry' too one day. I hope you will admire its clear and tidy beauty.

The second of the Big Three, Eratosthenes (c. 284–c. 192 BC), measured the circumference of the earth in a very simple but clever way – using geometry. He knew that on the Summer Solstice, the longest day of the year, the sun was directly overhead at a place called Syene. So he measured the angle of the sun on that day at Alexandria (where he was librarian of a famous museum and library), which was around 5,000 stades due north of Syene. (A 'stade' was a Greek measurement of distance, around a tenth of a modern mile.) From these measurements he used geometry to calculate that the earth is about 250,000 stades around. So, was he close? Eratosthenes' prediction of 25,000 miles is not very far off the actual 24,901.55 miles (around the equator) that we know today. Notice that Eratosthenes thought that the earth was round. The idea that the earth was a large flat surface and that people could sail off the edge was not always believed, despite the stories that are told about Christopher Columbus and his voyage to America.

The last of the Big Three also worked at Alexandria, the city in northern Egypt founded by Alexander the Great. Claudius Ptolemy (c. 100–c. 178), like many scientists of the ancient world, had very wide interests. He wrote about music, geography, and the nature and behaviour of light. But the work that brought him lasting fame is the *Almagest*, the title given to it by the Arabs. In this book, Ptolemy brought together and extended the observations of many Greek astronomers, including charts of the stars, calculations of the movements of the planets, moon, sun and stars, and the structure of the universe. He assumed, like everyone else at the time, that the earth is at the centre of everything, and that the sun, moon, planets and stars revolve around it in a circular fashion. Ptolemy was a very good mathematician, and found that by introducing a few corrections he was able to account for the movements of the planets that he, and many people before, had noticed.

It is quite difficult to explain the sun going around the earth when in fact the opposite happens. Ptolemy's book was essential reading for astronomers in the Islamic lands and in the European Middle Ages. It was one of the first works to be translated into Arabic, and then translated again into Latin, so much was it admired. In fact, Ptolemy was considered the equal of Hippocrates, Aristotle and Galen by many, although for us, these three get their own separate chapters.

CHAPTER 4

The Father of Medicine
HIPPOCRATES

The next time you have to see the doctor, ask if he or she took the Hippocratic Oath at their graduation ceremony. Not all modern medical schools require their students to recite it, but some do, and this oath, written more than 2,000 years ago, has something to say to us still. We shall see what that is shortly.

Even though Hippocrates' name is attached to this famous oath, he probably didn't write it. In fact, he wrote only a few of the sixty or so treatises (short books on specific topics) that bear his name. We know only a little about Hippocrates the man. He was born about 460 BC, on the island of Cos, not far from present-day Turkey. He practised as a doctor, taught medicine (for money) and probably had two sons and a son-in-law who all were doctors. There is a long history of medicine being a family tradition.

The Hippocratic Corpus (a *corpus* is a group of writings) was actually written by many individuals, over a long period of time, perhaps as long as 250 years. The various treatises in the Corpus argue different points of view, and they deal with lots of different matters. These include diagnosing and treating diseases, how to cope with broken bones and dislocated joints, epidemics, how to stay healthy, what to eat, and how the environment can influence our health. The treatises also help doctors know how to behave, both with their patients and with other doctors. In short, the Hippocratic writings cover just about the whole of medicine as it was practised at the time.

Just as remarkable as the range of subjects covered is how long ago the treatises were written. Hippocrates lived before Socrates, Plato and Aristotle, and on Cos, a small, remote island. It is amazing that anything written so long ago survives at all. There were no printing presses, and words had to be copied laboriously by hand on parchment, scrolls, clay and other surfaces, and then passed from person to person. Ink fades, wars lead to destruction, and insects and weather take their toll. We generally have only copies of those writings, made much later by generations of interested people. The more copies that were made, the greater the chance that some of them would survive.

The Hippocratic treatises laid the foundation of Western medicine, and therefore Hippocrates still occupies a special position. Three broad principles have guided medical practice for centuries. The first still underpins our own medicine and medical science: the firm belief that people fall ill because of 'natural' causes that have rational explanations. Before the Hippocratics, in Greece and its neighbouring lands, disease was assumed to have a supernatural dimension. We fall ill because we have offended the gods, or because someone with unearthly powers cast a spell on us, or is displeased with us. And if witches, magicians and gods caused disease, it was best to leave priests or magicians to figure out why the disease had happened and how best to cure it. Many people, even today, use magical remedies, and faith-healers are still with us.

The Hippocratics were not priest-healers, they were doctors, who believed that disease was a natural, normal event. One treatise, *On the Sacred Disease*, shows this very clearly. This short work is about epilepsy, a common disorder then as now: we think both

Alexander the Great and Julius Caesar suffered from the condition. People with epilepsy have fits, during which they can become unconscious and experience muscle-twitching, and their bodies twist about. Sometimes, they wet themselves. Gradually, the fit subsides and they regain control of their bodies and mental functions. Those who suffer from epilepsy nowadays look upon it as a 'normal', if inconvenient, episode. But seeing someone during an epileptic fit can be pretty disturbing, and so dramatic and mysterious were the seizures that the ancient Greeks assumed the condition had a divine cause. So they called it the 'Sacred Disease'.

The Hippocratic author of the treatise was having none of this. His famous opening sentence states bluntly, 'I do not believe that the "Sacred Disease" is any more divine or sacred than any other disease, but, on the contrary, has specific characteristics and a definite cause. Nevertheless, because it is completely different from other diseases, it has been regarded as a divine visitation by those who, being only human, view it with ignorance and astonishment.' The author's theory was that epilepsy is caused by a blockage of phlegm in the brain. Like most theories in science and medicine, better ones have replaced it. But the firm statement – that you can't say a disease has a supernatural cause simply because it is unusual or mysterious or hard to explain – might be said to be the guiding principle of science throughout the ages. We may not understand it now, but with patience and hard work, we can. This argument is one of the most lasting things handed down to us by the Hippocratics.

The second Hippocratic principle was that both health and diseases are caused by the 'humours' in our bodies. (An old expression is that someone is in a good or bad humour, meaning in a good or bad mood.) This idea is most clearly set out in the treatise *On the Nature of Man*, which might have been written by Hippocrates' son-in-law. Several other Hippocratic works mention two humours – phlegm and yellow bile – as the causes of disease. *On the Nature of Man* added two more: blood and black bile. The author argued that these four humours play essential roles in our health, and when they get out of balance (when there is too much or too little of one or the other) then disease occurs. You've probably seen your own bodily fluids when you've been ill. When we have a fever, we break out in a sweat; when we have a cold or chest infection, our noses run and we cough up phlegm. When we have upset tummies, we vomit, and diarrhoea expels fluids from the other end. A scrape or cut can cause the skin to bleed. Less common today is jaundice, when the skin turns yellow. Jaundice can be caused by many diseases affecting those organs that make the bodily fluids, including malaria, which was common in ancient Greece.

The Hippocratics associated each of these humours with an organ in the body: blood with the heart, yellow bile with the liver, black bile with the spleen, and phlegm with the brain. The author of *On the Sacred Disease* thought that epilepsy was caused by blocked phlegm in the brain. Other diseases, not just ones such as colds or diarrhoea with their obvious changes in fluids, were associated with changes in the humours. Each of the humours had its properties: blood is hot and moist; phlegm, cold and moist; yellow bile, hot and dry; black bile, cold and dry. These kinds of symptoms can actually be seen in those who are ill: when a wound is inflamed with blood, it's hot, and when we have a runny cold we *feel* cold and shiver. (Galen, who developed Hippocratic ideas about 600 years later, also gave these same characteristics of hot, cold, moist and dry to the foods we eat, or drugs we might take.)

The cure for all illnesses was to restore whatever balance of humours was best for each patient. That meant that in practice Hippocratic medicine was more complicated than simply following instructions to return each humour back to its 'natural' state. Each individual patient had his or her own healthy balance of the humours, so the doctor had to know all about his patient: where they lived, what they ate, how they earned their living. Only by knowing his patient well could he tell the patient what was likely to happen, that is, give them a *prognosis*. When we are sick, we want most of all to know what to expect, and how we might get better. Hippocratic doctors placed great store in being able to predict just what would happen. Getting that right increased their reputations and brought them more patients.

The medicine that the Hippocratics learned, and then taught to their pupils (often their sons or sons-in-law), was based on careful observation of diseases and the course they took. They wrote down their experiences, often in the form of short summaries called 'aphorisms'. *Aphorisms* was one of the Hippocratic works most widely used by later doctors.

The Hippocratics' third important approach to health and disease was summarised by the Latin phrase *vis medicatrix naturae*, which means 'the healing power of nature'. Hippocrates and his followers interpreted the movements of humours during disease as signs of the body's attempt to heal itself. So sweating, bringing up phlegm, vomiting and the pus of abscesses were viewed as the body expelling – or 'cooking' (they used kitchen metaphors a lot) – the humours. The body did this to get rid of excesses or modify or purify bad humours that had been changed by disease. The doctor's job was therefore to assist nature in the natural healing process. The doctor was nature's servant, not her master, and the processes of disease were to be learned by close observation of exactly what occurred during disease. Much later, one doctor coined the phrase 'self-limited disease' to describe this tendency, and we all know that many illnesses get better by themselves. Doctors sometimes joke among themselves that if they treat a disease it will be gone in a week, but if they don't it will take seven days. The Hippocratics would have agreed.

Besides their many works on medicine and surgery, hygiene and epidemics, the Hippocratics left us the Oath, still a source of inspiration to doctors today. Some of this short document is concerned with the relationships between the young student and his master, and between doctors. Much of it, however, deals with the appropriate behaviour that doctors ought to adopt with their patients. They ought never to take advantage of their patients, gossip about secrets they might hear from the sick, or administer a poison. All these issues are still vital in medical ethics today, but one Hippocratic statement in the Oath seems particularly timeless: *I will use my power to help the sick to the best of my ability and judgement; I will abstain from harming or wrongdoing any man by it.* 'To do the sick no harm' ought still to be every doctor's aim.

CHAPTER 5

'The Master of Those Who Know'
ARISTOTLE

'All men by nature desire to know,' said Aristotle. You have probably met someone like this, always keen to learn more. Perhaps you've also come across know-it-alls who have lost the curiosity that always remained important to Aristotle. His hopeful view was that people will strive for knowledge about themselves and the world. We know, unfortunately, that this isn't always the case.

Aristotle spent his whole life learning and teaching. He was born in 384 BC, in Stagira, Thrace (now Khalkidhiki in Greece). He was the son of a doctor, but from the age of about ten, he was looked after and taught by his guardian Proxenus. When he was about seventeen, Aristotle went to Athens to study at Plato's famous Academy. He stayed there for twenty years. Although Aristotle's approach to the natural world was completely different from Plato's, Aristotle was very fond of his teacher and wrote about his work lovingly after Plato's death in 347 BC. Some say that the history of Western philosophy is a series of footnotes to Plato; what this means is that Plato raised many of the questions that philosophers still think about. What is the nature of beauty? What is truth, or knowledge? How can we be good? How can we best organise our societies? Who makes the rules we live by? What does our experience of the things of the world tell us about what they 'really' are?

Aristotle, too, was intrigued by many of these philosophical questions, but he tended to answer them in a way we might call 'scientific'. He was, like Plato, a philosopher, but he was a *natural* philosopher, what we are calling a 'scientist'. The branch of philosophy that most excited him was logic – how we can think more clearly. He was always busy with the world about him, on the ground and in the skies, and with the way natural things change.

Much of what Aristotle wrote has been lost, but we are lucky to have some of his lecture notes. He left Athens after Plato died, probably because he felt unsafe as a foreigner there. He spent some years in the city of Assos (now in Turkey), where he set up a school, married the daughter of the local ruler and, after she died, lived with a slave girl with whom he had a son, Nicomachos. It was here that Aristotle began his biological investigations, which he continued on the island of Lesbos. In 343 BC, Aristotle took a very important job: tutor to Alexander the Great, in Macedonia (now a separate country just north of Greece). He hoped to turn his pupil into a philosophically sensitive ruler; he didn't succeed, but Alexander came to rule over much of the known world, including Athens, so Aristotle could safely return to that city. Instead of going back to Plato's Academy, Aristotle founded a new school just outside Athens. It had a public walkway (*peripatos* in Greek), so Aristotle's followers became known as Peripatetics, or those constantly moving around: an appropriate name considering how much Aristotle himself moved from place to place. After Alexander's death, Aristotle lost his support in Athens, so he moved one last time, to Chalcis, in Greece, where he died shortly afterwards.

Aristotle would have been puzzled to be described as a scientist; he was simply a philosopher in the literal meaning of the word: a lover of wisdom. But he spent his life trying to make sense of the world around him, and in ways that we would now describe as scientific. His vision of the earth, its creatures and the heavens around it, influenced our understanding for more than 1,500 years. Along with Galen, he towered over all other ancient thinkers. He built on what had gone before, of course, but he was no armchair philosopher. He actually engaged with the material world as he attempted to understand it.

We can separate his science into three parts: the living world (plants and animals, including human beings); the nature of change, or movement, much of which is contained in a work of his entitled *Physics*; and the structure of the heavens, or the relationship of the earth to the sun, moon, stars and other heavenly bodies.

Aristotle spent much time studying how plants and animals are put together and how they work. He wanted to know how they develop before birth, hatching, or germination, and then how they grow. He had no microscope, but his eyesight was obviously good. He described brilliantly the way chicks develop in an egg. After a batch of eggs had been laid, he cracked one each day. The first sign of life he saw was a tiny speck of blood pulsating in what would become the chick's heart. This convinced him that the heart was the key organ in animals. He believed the heart was the centre of emotion and what we would call mental life. Plato (and the Hippocratics) had located these psychological functions in the brain, and they were correct. Nevertheless, when we are frightened, or nervous, or in love, our hearts beat faster, so Aristotle's theory was not silly. He attributed the functions of higher animals, such as human beings, to the activities of a 'soul', which has various faculties, or functions. In humans, there were six main faculties of the soul: nutrition and reproduction, sensation, desire, movement, imagination, and reason.

All living beings have some of these capacities. Plants, for instance, can grow and reproduce; insects such as ants can also move and feel. Other bigger and more intelligent animals acquire more functions, but Aristotle believed that only human beings could reason – that is, they could think, analyse and decide on a course of action. Human beings therefore sat at the top of Aristotle's *scala naturae* ('scale of nature', or 'chain of being'). This was a kind of ladder upon which all living things could be arranged, beginning with simple plants and working upwards. This idea was taken up again and again by different naturalists, people who study nature, especially plants and animals. Look out for it in later chapters.

Aristotle had a good way of working out what is done by the various parts of a plant or animal, such as the leaves, wings, stomach or kidneys. He assumed that the structure of each part was designed with a particular function in mind. Thus, wings were designed for flight, stomachs for the digestion of food, and kidneys for the processing of urine. This kind of reasoning is called *teleo-logical*: a *telos* is a final cause, and this way of thinking focuses on what things are like or what they do. Think about a cup, or a pair of shoes. They both have the shape they have because the person who made them had a specific purpose in mind: to hold liquids for drinking, and to protect feet while walking. Teleological reasoning will appear later in the book, not just in explaining why plants or animals have the various parts that they do, but in the wider physical world as well.

Plants germinate and animals are born, they grow and then die. The seasons regularly come and go. If you drop something, it falls to the ground. Aristotle wanted to explain changes like these. Two ideas were very important to him: 'potentiality' and 'actuality'. Teachers or parents may tell you to reach your potential: that usually means something like getting the best possible marks in a test, or running a race as fast as you can. That is part of Aristotle's idea, but he saw a different kind of potential in things. In his view, a pile of bricks has the potential to become a house, and a lump of stone has the potential to be a statue. Building and sculpting trans-form these inanimate objects from a kind of potential to a kind of finished thing, or 'actuality'. Actuality is an end-point of potenti-ality, when things with potentiality find their 'natural state'. For example, when things fall, like apples from an apple tree, Aristotle thought that they seek their 'natural' state, which is on the earth. An apple will not suddenly sprout wings and fly, because it and all other things in our world seek the earth, and a flying apple would be very unnatural. That fallen apple may continue to change – it will rot, if no one picks it up and eats it, because that is also part of an apple's cycle of growth and decay. But just by falling it has achieved a kind of actuality. Even birds return to earth after they soar into the sky.

If the 'natural' resting place of things is on the firm earth, what about the moon, the sun, planets and stars? They may be up there, like an apple hanging in a tree, or a boulder on a mountain ledge, but they never come crashing to earth. Good thing, too. Aristotle's answer was simple. From the moon downwards, change is always happening; this is because the world is composed of the four elements: fire, air, earth and water (and their properties: hot and dry fire, hot and moist air, cold and dry earth, and cold and moist water). But above the moon, things are made instead of a fifth, unchanging element, the *quintessence* ('fifth essence'). The heavenly bodies move forever in perfect circular motion. Aristotle's universe filled a fixed space but not a fixed time. The sun, moon and stars have been moving for all eternity around the earth, which floats at the centre of it all. There is a lovely paradox here, for the earth, the centre, is also the only part of the universe in which change and decay can take place.

What caused all this movement around the earth in the first place? Aristotle was very concerned with *cause*. He developed a scheme to try to explain causes by breaking them down into four kinds. These were called material, formal, efficient and final causes, and he thought that human activities, as well as what happens in the world, could be broken down and understood this way. Think about making a statue from a lump of stone. The stone itself is the *material* cause, the matter out of which it is made. The person making the statue arranges things in a certain, *formal* manner, so that the statue takes shape. The *efficient* cause is the act of chiselling against the stone to make the shape. The *final* cause is the idea that the sculptor had in mind – the shape, say, of a dog or horse – which was the plan of the whole activity to begin with.

Science has always dealt with causes. Scientists want to know what happens and why. What causes a cell to start dividing endlessly, with the result that a person develops a cancer? What turns leaves brown, yellow and red in the autumn, when they have been green all summer? Why does bread rise up when you put yeast into it? These and many similar questions can be answered in terms of various 'causes'. Sometimes the answers are pretty simple; sometimes they are very complicated. Mostly, scientists deal with what Aristotle called efficient causes, but the material and formal causes are also important. Final causes raise a different set of issues. In scientific experiments today, scientists are content with explaining the processes rather than seeking any larger explanation or final cause, which has more to do with religion or philosophy.

Back in the fourth century BC, however, Aristotle believed that these final causes were part of the picture. Looking at the universe as a whole, he argued that there must be some final cause that started off the whole process of movement. He called this the 'unmoved mover', and later many religions (Christianity, Judaism and Islam, for example) identified this force with their God. This was one reason why Aristotle continued to be celebrated as such a powerful thinker. He created a world-view that dominated science for almost 2,000 years.

CHAPTER 6

The Emperor's Doctor
GALEN

Galen (129–c. 210) was very clever and was not afraid to say so. He scribbled constantly, and his writings are full of his own opinions and accomplishments. More of his words survive than those of any other author from ancient times, which proves that people valued Galen's works very highly. There are twenty fat volumes that you can read, and he actually wrote many more. So we know more about Galen than we do about most other ancient thinkers. It doesn't hurt that Galen also adored writing about himself.

Galen was born in Pergamum, now part of Turkey but then on the fringes of the Roman Empire. His father was a prosperous architect who was devoted to his gifted son, providing him with a sound education (in Greek) which included philosophy and mathematics. Who knows what might have happened had his father not had a powerful dream, telling him that his son ought to become a doctor? Galen changed his studies to medicine. After his father's death left him well-off, he spent several years travelling and learning, spending time at the famous library and museum in Alexandria in Egypt.

Back in Pergamum, Galen became a doctor to the gladiators – the men chosen to entertain well-to-do citizens by fighting each other, or by facing lions and other beasts in the arena. Taking care of them was an important job, since the poor men needed to be patched up between the shows so they could keep on fighting. By his own account, Galen was extremely successful. He would have had dramatic experience in the surgical treatment of wounds. He also acquired a considerable reputation among the rich and, around AD 160, he took himself to Rome, the capital of the Roman Empire. He began writing on anatomy (the study of the bodily structures of humans and animals) and physiology (the study of what those structures do). He also went on a military campaign with the Emperor Marcus Aurelius. The emperor was the author of a famous series of *Meditations* and the two men discussed philosophy during the long campaign. Marcus Aurelius appreciated Galen, and Galen profited from the emperor's support. A steady stream of important patients were sent his way whom, if Galen's reports are to be believed, he always cured if they could be.

Galen's medical hero was Hippocrates, even though he had been dead for more than 500 years. Galen saw himself as completing and extending the master's legacy, and, in many ways, this is exactly what he did. He produced commentaries on many of the Hippocratic works, and assumed that the works that agreed most closely with his own views were by Hippocrates himself. His comments on Hippocrates are still valuable, not least because Galen was an expert linguist with a good eye for the changing meanings of words. Most importantly, he put the Hippocratic doctrine of the humours in the form that was used for more than a thousand years. Imagine being that influential!

The idea of the balance and imbalance of the humours was central to Galen's medical practice. Like Hippocrates, he believed that the four humours – blood, yellow bile, black bile and phlegm – were, in special ways, hot or cold, and moist or dry. To treat a malady, you chose an 'opposite' remedy, but also one of the same intensity. So diseases that were hot and moist in the third degree, for example, would be treated with a remedy that was cold and dry in the third degree. For example, if the patient had a runny nose and felt chilled, drying, warming medicines and food would be used. By rebalancing the humours, you could restore a healthy 'neutral' state. This was all very logical and simple, but in reality things were more complicated. Doctors still needed to know a great deal about their patients, and administer their remedies with care. Galen was always quick to point out when other doctors got it wrong (which was often) so that everyone knew his diagnoses and therapies were better. He was a shrewd doctor, much in demand, who paid great attention to the mental as well as physical aspects of health and disease. He once diagnosed a case of 'love-sickness', where a young lady became weak and nervous whenever a handsome male dancer was performing in town.

Galen came up with the practice of feeling his patient's pulse – something that doctors still do. He wrote a treatise on how the pulse – slow or fast, strong or faint, regular or irregular – could be useful in diagnosing disease, even though he had no idea about the circulation of the blood.

Galen was more interested in anatomy than the Hippocratics. He opened up the bodies of dead animals and examined human skeletons wherever he could. Dissecting human bodies was frowned upon in ancient societies, so Galen could not do that, although we think that a few earlier doctors might have been allowed to examine the bodies of condemned criminals while they were still alive. Galen learned about human anatomy from dissections of animals, like pigs and monkeys, and by lucky chances – the discovery of a decaying dead body, or bad injuries that showed the structure of skin, muscle and bone. Scientists still use animals in their research, but they must be careful to be clear about where they got their information. Galen often forgot to mention where he had got his facts from, so it could be confusing.

Anatomy was, for Galen, an important subject in its own right, but it was also fundamental for understanding what the organs of the body actually do. One of his most influential treatises was called *On the Uses of the Parts*, which looked at the structures of the 'parts', or organs, and what role they played in the working of the whole human body. Galen assumed, as you would do, that each part does do something, otherwise it wouldn't be there. (I doubt if he ever saw the human appendix. That tiny part of our digestive organs probably long, long ago helped us to digest plants, but it doesn't have a function any more.)

At the centre of all bodily function was a substance the Greeks called *pneuma*. 'Pneuma' is not easily translated into English: we'll use 'spirit' but it also has the idea of 'air'; it has given rise to various medical terms in our own times, such as 'pneumonia'. For Galen, the body contained three kinds of pneuma, and understanding what they each did was central to understanding how the body functions. The most basic kind of pneuma was associated with the liver, and was concerned with nutrition. The liver, Galen believed, was able to draw material from the stomach after it had been eaten and digested, turn it into blood and then infuse it with 'natural' spirit. This blood from the liver then coursed through the veins throughout the body, to nourish the muscles and other organs.

Some of this blood passed from the liver through a large vein, the *vena cava*, into the heart where it was further refined with another spirit, the 'vital' one. The heart and lungs worked together in this process, some of the blood passing through the pulmonary artery (going from the right side of the heart) into the lungs. There it nourished the lungs and also mixed with the air we breathe in through the lungs. Meanwhile, some of the blood in the heart passed from right to left through the middle portion of the heart (the septum). This blood was bright red because, Galen thought, it had the vital spirit infused within it. (Galen recognised that blood in the arteries is a different colour from blood in the veins.) From the left side of the heart, blood went out via the aorta, the large artery taking blood from the left chamber, or ventricle, of the heart, in order to warm the body. Despite his appreciation of the importance of blood in the life of an individual, Galen had no sense that the blood *circulates*, as William Harvey was to discover almost 1,500 years later.

In Galen's scheme, some of the blood from the heart also went to the brain, where it was mixed with the third kind of pneuma, the 'animal' spirit. This was the most refined kind of spirit. It gave the brain its own special functions as well as flowing out through the nerves, enabling us to move using our muscles and to experience the external world using our senses.

Galen's three-part system of spirits, each associated with the important organs (liver, heart, brain), was accepted for more than a thousand years. It's worth remembering that Galen used this system primarily to explain how our bodies work when we are healthy. When he tended sick patients, he continued to rely on the system of humours devised by the Hippocratics.

Galen also wrote about most other aspects of medicine, such as drugs and their properties, the diseases of the special organs like the lungs, hygiene, or how to preserve health, and the relationship between our minds and our bodies. His thinking was very sophisticated. In fact he believed that a doctor should be both a philosopher and an investigator: a thinker and an experimenter. He argued that medicine should, above all, be a rational science, and he paid a lot of attention to the best ways to gain good, reliable knowledge. Later doctors, who also saw themselves as learned men of science, liked Galen's mix of practical advice (based on his vast experience) and broad thinking. No single Western doctor in all history has exerted such an influence for so long.

There are several reasons for Galen's long shadow. First, he had a very high opinion of Aristotle, so that the two were often spoken about together. Like Aristotle, Galen was a deep thinker and an energetic investigator of the world. Both believed this world had been designed, and praised the Designer. Galen was not a Christian, but he believed in a single God, and it was easy for early Christian commentators to include him in the Christian fold. His confidence meant that he had an answer for everything. Like most people who write many books over a long period, he was not always consistent, but he was always definite in his opinions. He was commonly referred to later as 'the divine Galen', a label of which he would have been proud.

CHAPTER 7

Science in Islam

Galen did not live to see the decline of the Roman Empire, but by AD 307 it had been split in two. The new emperor, Constantine (280–337), moved his seat of power to the east – to Constantinople, now Istanbul in modern Turkey. There he would be nearer to the eastern part of the Empire, lands that we now call the Middle East. The learning and wisdom contained in the Greek and Latin manuscripts, as well as the scholars who were able to study them, began to move eastwards.

A new religion arose in the Middle East: Islam, which followed the teachings of the great prophet Muhammad (570–632). Islam would come to dominate most of the Middle East and North Africa, and even as far as Spain and East Asia, but in the two centuries after Muhammad's death, the new religion was largely confined to Baghdad and other settlements in the area. All Muslim scholars studied the Qur'an, the central religious text of Islam. Yet many of them were also interested in the many manuscripts that had been brought there after Rome was attacked in 455. A 'House of Wisdom' was established in Baghdad, which encouraged ambitious young men to join in the translation and study of these old manuscripts.

Many of the old manuscripts were still in the original Greek or Latin, but others had already been translated into Middle Eastern languages. The works of Aristotle, Euclid, Galen and other thinkers of ancient Greece were all translated – a very good thing too, as some of the original versions have since disappeared. Without Islamic scholars, we wouldn't know half as much as we do about our scientific ancestors. And more than that: it was their translations that formed the foundation of European science and philosophy after about 1100.

Islamic science straddled East and West, just as the Muslim lands did. Aristotle and Galen were just as admired in Islamic lands as they were in Europe; Aristotle made his way into Islamic philosophy, and Galen became the master of medical theory and practice. Meanwhile, ideas from India and China were introduced to the West. Paper from China made it much easier to produce manuscripts, though they still had to be copied by hand, and mistakes were common. From India came the numerals 1 to 9, the idea of 0, and place-holding, all invented by Indian mathematicians. Europeans could do calculations using Roman numerals, such as I, II and III, but it was difficult, even if that was what they were used to. It's simpler to use 4 × 12 than IV × XII, isn't it? When Europeans translated Islamic works into Latin, they called these numerals 'Arabic' – strictly speaking, they should have said 'Indian-Arabic', but what a mouthful! The word 'algebra' actually comes from the term *al-jabr*, in the title of a widely-translated book by a ninth-century Arab mathematician. There is more about algebra in Chapter 14.

Islamic scholars made many significant discoveries and observations. If you have ever climbed up a mountain, or gone to a country that is high above sea level, you might know that breathing is more difficult because the air is thinner. But how high would you have to go before you couldn't breathe any more? In other words, how high is the atmosphere, the band of breathable air that surrounds the globe? Ibn Mu'adh, in the eleventh century, hit upon a smart way of finding out. He reasoned that twilight – that is, when the sun has set, but the sky is still light – happens because the sun's dying rays are being reflected by water vapour high in the atmosphere. (Many Islamic scholars were interested in such tricks of the light.) Observing how fast the sun had disappeared from the evening sky, he worked out that the sun at twilight was 19 degrees below the horizon. From there, he calculated that the height of the atmosphere was fifty-two miles – not so far off the height of sixty-two miles we now think is correct. Simple, but very impressive.

Other Islamic scholars investigated the reflection of light in a mirror, or the strange effect of light passing through water. (Put a pencil in a half-filled glass of water: it looks bent, doesn't it?) Most Greek philosophers had assumed that seeing something involved light coming out of the eye, hitting the object that was being viewed, and bouncing back. Islamic scientists mostly favoured the more modern view, that the eye receives light from the things we see, which the brain then interprets. Otherwise, as they pointed out, how is it that we can't see in the dark?

Many in the Middle East did see in the dark: their astronomers looked at the stars, and their charts and tables of the night skies were better than those of Western astronomers. They still thought that the earth was the centre of the universe, but two Islamic astronomers, al-Tusi in Persia and Ibn al-Shatir in Syria produced diagrams and some calculations that were important to the Polish astronomer Copernicus 300 years later.

Medicine, more than any other Islamic science, had the greatest impact on European thinking. Hippocrates, Galen and the other Greek doctors were lovingly translated and commented on, but several Islamic doctors also made names for themselves. Rhazes (c. 854–c. 925), as he is known in the West, wrote important works in several subjects besides medicine; he also left an accurate description of smallpox, a much-feared disease, which often killed its victims and scarred those who survived. Rhazes distinguished smallpox from measles, which is still a disease that children and some adults catch. Like smallpox, measles produces a rash and fever. Smallpox is now happily extinct, the result of an international campaign to protect people by vaccination, led by the World Health Organization (WHO). The last case occurred in 1977: Rhazes would have been pleased.

Avicenna (980–1037) was the most influential Islamic doctor. Like many other eminent Islamic scholars, he was busy in many fields: not just medicine, but also philosophy, mathematics and physics. As a scientist, Avicenna developed Aristotle's views on light, and corrected Galen on a number of points. His *Canon of Medicine* was one of the first books in Arabic that was translated into Latin, and it was used as a textbook in European medical schools for almost 400 years. It is still used in some modern Islamic countries, which is unfortunate, since it is sadly out of date now.

For more than 300 years, the most important scientific and philosophical work was done in Islamic countries. While Europe slept, the Middle East (and Islamic Spain) was busy. The most important places were Baghdad, Damascus, Cairo, and Cordoba (in Spain). These cities all shared one characteristic: enlightened rulers who valued and even funded research, and were tolerant of scholars of all faiths. Thus, Christians and Jews as well as Muslims contributed to this movement. Not all Islamic rulers were happy for knowledge to be gained from whatever source; some held that the Qur'an contained everything a person needed to know. These tensions continue today. Science has always been strongest in cultures that are open to the new, since finding out about the world can produce surprises.

CHAPTER 8

Out of the Darkness

We expect scientists to be trying to discover new things, and for science to be constantly changing. But what would science be like if we thought that everything had already been discovered? Being a top scientist might then involve just reading about other people's discoveries.

In Europe, this backward-looking view became the norm after the fall of the Roman Empire in AD 476. By then, Christianity had become the official religion of the Empire (Constantine had been the first emperor to convert to Christianity), and only one book mattered: the Bible. St Augustine (354–430), one of the most influential early Christian thinkers, had put it this way: 'The truth is rather in what God reveals than in what groping men surmise'. There was no room for those scientists who were 'groping' for knowledge; the ancients had already discovered everything worth knowing in science and medicine. Besides, it was far more important to focus on getting to Heaven and avoiding Hell. Being a 'scientist' might mean just studying Aristotle and Galen. And for 500 years, from about AD 500 to 1,000, even that was difficult, since very few Greek and Latin texts from the classical world were available. Nor did very many people know how to read.

The Germanic tribes who sacked Rome in 455 did bring some useful things with them, however. Wearing trousers instead of togas was one (though women had to wait a while longer). So were new grain crops such as barley and rye, and eating butter instead of olive oil. There were technological innovations in that 'dark' half-millennium, too: it saw new ways of growing crops and of ploughing the land. Building churches and cathedrals encouraged craftsmen and architects to experiment with new styles, and find better ways of spreading the heavy weight of stone and timber. This meant they could build ever-bigger and grander cathedrals, and some of these early buildings still take your breath away. They are reminders that what is sometimes called the 'Dark Ages' was not without its light.

With the coming of the second millennium of the Christian era, however, the pace of discovery picked up. St Thomas Aquinas (c. 1225–74) was the greatest medieval theologian. He admired Aristotle immensely, and he meshed Christian thought with Aristotelian science and philosophy. Aristotle, together with Galen, Ptolemy and Euclid, shaped the medieval mind. Their writings needed to be translated, edited and commented upon. Originally much of this activity took place in monasteries, but gradually it moved to the universities, which were first introduced in this period.

The Greeks had had schools: Aristotle studied at his teacher Plato's Academy, and established his own school in turn. The House of Wisdom in Baghdad was also a place where people came together to study and learn. But the new universities of Europe were different, and most of them have survived to this day. Many were established by the Church, but community pride and rich supporters helped some towns and cities start their own university. The Pope authorised the foundation of several universities in southern Italy. The University of Bologna (established around

1180) was the first to open its doors, but, within a century or so, there were universities at Padua, Montpellier, Paris, Cologne, Oxford and Cambridge. The name 'university' comes from the Latin word meaning 'whole', and these institutions were supposed to cover the whole of human knowledge. They usually had four schools, or 'faculties': Theology, of course (Aquinas called theology 'the queen of the sciences'), Law, Medicine and Arts. The medical faculties initially relied mostly on Galen and Avicenna. Medical students also studied astrology, because of the widespread belief in the power of the stars to affect humans, for better or worse. Mathematics and astronomy – which we would think of as very scientific – were generally taught in the arts faculty. Aristotle's vast works were studied in all the faculties.

Many of the 'scientists' of the Middle Ages were either doctors or clergymen, and most of them worked at the new universities. The faculties of medicine gave their graduates degrees – Doctor of Medicine (MD) or Bachelor of Medicine (MB) – and this in turn separated these physicians from the surgeons, apothecaries (pharmacists) and other medical practitioners who learned their trades in other ways. Their university education didn't necessarily make doctors more interested in finding out new things (they preferred to rely on Galen, Avicenna and Hippocrates). But from around 1300, anatomy teachers began to dissect bodies to show the internal organs to their students, and autopsies were sometimes carried out on royalty, or when the death was suspicious (or both). None of this necessarily made doctors more able to treat diseases, especially those that swept through communities.

What we now call the Black Death, a form of plague, entered Europe for the first time in the 1340s. It probably came from Asia, along trade routes, and killed about one third of the people of Europe in the three years it took to make its rounds. As if that were not enough, it returned ten years later, and then with depressing regularity for the next 400 years. Some communities established special hospitals for plague sufferers (hospitals, like universities, are a medieval gift to us), and Boards of Health were set up in some places. The plague also led to the use of quarantine in cases of disease thought to be contagious. 'Quarantine' comes from the number 40 (in Venetian, *quaranta*), which was the number of days that the sick or suspected person was placed in isolation. If the individual recovered in that time, or showed no signs of the disease, he or she could be released. The playwright William Shakespeare was born in Stratford-upon-Avon in a plague year in England (1564), and his career was interrupted several times, when plague epidemics forced the theatres to close down. Shakespeare has Mercutio, in *Romeo and Juliet*, say 'A plague on both your houses!', to condemn the two warring families. His audience would have understood what he meant. Most doctors thought that plague was a new disease, or at least one that Galen had not written about, and so they had to cope without his advice: remedies included blood-letting and drugs that would make the patient vomit or sweat, popular cures for other diseases at the time. Galen didn't know everything, after all.

Neither, it seemed, did Aristotle. His ideas about why something moves through the air were widely discussed by Roger Bacon (*c.* 1214–94) at the University of Oxford, by Jean Buridan (*c.* 1295–*c.* 1358) at the University of Paris, and several others. It was called the 'impetus problem' and needed to be solved. Take the example of a bow and arrow. The arrow flies because we pull back the bow's string and quickly release it, pushing the arrow through the air. We have applied a force and given it *momentum* (a concept that we'll talk more about later). Bacon and Buridan called this 'impetus', and they realised that Aristotle did not have a correct explanation for the fact that the further we pull back the bow string, the further the arrow will fly. Aristotle said that an apple will fall to the earth because that is its 'natural' resting place. The arrow will eventually come to earth, too, and yet Aristotle had said it moved only because there was a force behind it. So, if there was a force when the arrow left the string, why did the force seem to wear out?

These and similar problems made some people think that Aristotle hadn't got everything correct. Nicolas Oresme (*c.* 1320–82), a churchman working in Paris, Rouen and elsewhere in France, wondered again about day and night. Rather than the sun racing around the earth every twenty-four hours, perhaps, he thought, the earth itself rotates on its axis over the course of a day. Oresme didn't challenge Aristotle's belief that the earth was at the centre of the universe, or that the sun and planets revolved around the earth. But perhaps that was a very slow journey (maybe it took the sun a year to make it around!), while the earth, at the centre of the universe, was spinning like a top.

These ideas were new, but 700 years ago people didn't necessarily think that new ideas were always good. Instead, they liked systems that were neat, tidy and complete. This is one reason why so many scholars wrote what we now call 'encyclopaedias': big works that took the works of Aristotle and the other ancient masters, and put them together – synthesising them – into gigantic wholes. 'A place for everything, and everything in its place': that could be the motto for this period. But trying to find that place for everything led some to realise that there were still puzzles to be solved.

CHAPTER 9

Searching for the Philosopher's Stone

If you could turn your aluminium Coca-Cola can into gold, would you? You probably would, but if everybody could do it, it wouldn't be quite so amazing, since gold would become common and not worth much. The old Greek myth of King Midas, who was granted his wish that everything he touched would turn to gold, reminds us that he wasn't being very clever. He couldn't even eat his breakfast, since his bread became gold as soon as he touched it!

King Midas was not alone in thinking that gold is special. Humans have always valued it, partly because of its wonderful feel and colour, partly because it is scarce, and only kings and other rich people possessed it. If you could discover how to make gold from more common substances – from iron or lead, for instance, or even from silver – your fame and fortune would be sealed.

Making gold in this way was one of the aims of a kind of early science called alchemy. Drop the 'al' from alchemy and you get a version of 'chemistry', and in fact the two are related, although these days we wouldn't call alchemy – with its dark connections to magic and religious belief – a science. However, in the past it was a thoroughly respectable activity. In his spare time, Isaac Newton (Chapter 16) dabbled in alchemy, buying a lot of scales, strangely shaped glassware, and other equipment. In other words, he set up a chemistry laboratory.

You might have been in a laboratory, or at least seen them in pictures or films; the name simply means places where you 'labour', or work. Long ago, laboratories were where alchemists worked. Alchemy has a long history, stretching back to ancient Egypt, China and Persia. The aim of alchemists was not always simply to change less valuable ('base') metals into gold: it was also to exert power over nature, to be able to control the things that surround us. Alchemy often involved the use of magic: saying spells, or making sure you did things in exactly the correct order. The alchemist experimented with substances, to see what happened when two were mixed together, or heated. Alchemists liked to work with things that had violent reactions, like phosphorous or mercury. It could be dangerous, but imagine the rewards if you managed to find just the right combination of ingredients to make the 'philosopher's stone'. This 'stone' (it would actually be some kind of special chemical) would then turn lead or tin into gold, or help you live forever. Just like in Harry Potter.

Harry Potter's adventures are fun, but they take place in a world of the imagination. The kinds of powers that the real magicians and alchemists dreamed of are not available in ordinary life either – even the life of the alchemist, and a lot of alchemists were tricksters, pretending to do things that they could not. But many others were honest workers who lived in a world in which everything seemed possible. In the course of their studies, they found out a lot about what we now call chemistry. They learned about distillation, for instance, the art of heating a mixture and collecting the substances that the mixture leaves behind at different times. Strong alcoholic drinks like brandy and gin are produced by distillation, which concentrates the alcohol. We call them 'spirits', a word we also use for ghosts, and for ourselves when we are being lively, or 'spirited'. It's a word that comes from the Latin *spiritus*, meaning 'breath' as well as 'spirit'. It also comes in part from alchemy.

Most people used to believe in magic (and some still do). Many famous scholars in the past also used their studies of the secrets of nature to uncover magical forces. One remarkable man thought he had the power to change the whole practice of science and medicine. His full name is a mouthful: Theophrastus Philippus Aureolus Bombastus von Hohenheim. Try saying that name fast, and you might understand why he would want to change it to the one we know him by: Paracelsus.

Paracelsus (*c.* 1493–1541) was born in Einsiedeln, a small town in the Swiss mountains. His father was a doctor and taught him about the natural world, about mining, and minerals, botany and medicine. He was raised as a Roman Catholic, but he grew up in the days of Martin Luther and the Protestant Reformation, and he had many Protestant friends and supporters, as well as Roman Catholic ones. He also made many enemies. He studied with several important churchmen, and although Paracelsus was always deeply religious, his faith, like everything else about him, was unique: it was based on chemistry.

Paracelsus studied medicine in Italy, and was always restless, moving from place to place. He travelled all over Europe, perhaps went to England, and was certainly in North Africa. He worked as a surgeon and an ordinary doctor, treated many rich and powerful patients, and seems to have been successful. However, he never looked as if he had any money and was always badly dressed. He liked to drink in bars or pubs with ordinary rather than posh people, and his enemies said he was addicted to alcohol.

Paracelsus had only one formal job, at the university in Basel, in his native Switzerland. He insisted on lecturing in German, instead of Latin, as all the other professors did, and one of the first things he did was to burn the works of Galen in the marketplace. He had no need for Galen, Hippocrates or Aristotle. He wanted to start over again. He was sure that his view of the universe was the correct one, and it was unlike any that had gone before.

Shortly after his bonfire, he was forced to leave town to continue his wanderings, staying a few months here, perhaps a year or so there, but always restless and ready to pack his few things and try somewhere else. He would take his manuscripts and chemical apparatus, and probably little else. Travel was always slow, on foot, on horseback or in a cart, along roads that were often muddy and dangerous. Given his way of life, it is amazing that he accomplished anything at all. In fact, while treating many patients, he also wrote many books, looked at the world around him, and was always doing chemical experiments.

Chemistry was his passion. When he said he didn't need the works of the ancients to guide his own studies, he meant it. He had no time for the four elements of air, earth, fire and water. Instead, for him, there were three basic 'principles' – salt, sulphur and mercury – into which everything could ultimately be separated. Salt gives things their shape, or solidness; sulphur is the reason why things can burn; and mercury is responsible for a thing's smoky or fluid state. Paracelsus interpreted the experiments in his laboratory by these three principles. He was interested in how acids can dissolve things, and how alcohol can be frozen. He burned substances and carefully examined what was left. He distilled many liquids and collected what was given off, as well as noting what was left behind. In short, he spent a lot of time in his laboratory, seeking to master nature.

Paracelsus believed that his chemical experiments would help him understand how the world works, and that chemistry would be the source of many new treatments for disease. Before him, most drugs that doctors used came from plants, and although Paracelsus sometimes used herbal remedies in his own medical practice, he preferred to give his patients medicines that he had studied in his laboratory. Mercury was a particular favourite of his. Mercury is actually very poisonous, but Paracelsus used it as an ointment for skin diseases, and believed it was the best remedy for a disease that had become common throughout Europe. This was syphilis, a disease that is usually spread by sexual contact, which causes horrible rashes on the skin, destroys people's noses, and usually kills them. An epidemic of syphilis broke out in Italy in the 1490s, around the time of Paracelsus's birth, killing many people. By the time he was a doctor, syphilis was so widespread that almost all doctors would have seen patients with it (and more than a few doctors suffered from it themselves). Paracelsus wrote on this new disease, describing many of its symptoms and recommending mercury to treat it. Although mercury could make your teeth fall out and your breath smell horribly, it got rid of the rash, so doctors used it for many years to treat syphilis and other diseases that caused rashes.

Paracelsus described many other diseases. He wrote about the injuries and illnesses suffered by those who worked down the mines, especially diseases of the lungs caused by horrible working conditions and long hours. Paracelsus's concern for lowly miners reflected his life spent among the ordinary people.

Hippocrates, Galen and other doctors before Paracelsus thought disease was the result of an imbalance within the body. For Paracelsus, however, disease resulted from a force that was outside the body. This 'thing' (he called it an *ens*, a Latin word which means a 'being' or 'substance') attacks the body, causes us to fall sick, and creates the kind of changes that doctors look for as clues to understand what the disease is. The *ens* could be a pimple or abscess, or a stone in the kidney. The important breakthrough that Paracelsus made was separating the patient and the disease. This way of thinking came into its own much later with the discovery of germs.

Paracelsus wanted to start science and medicine on the new foundations he provided. He said again and again that people ought not to read books but to see and experiment for themselves. He did, of course, want others to read the books he himself wrote, some of which were not published until after he died. His real message was 'Don't bother to read Galen, read Paracelsus.' His world was full of magical forces, which he believed he could understand and tame in the service of his science and medicine. His own alchemical dream was not just turning base metals into gold; rather, he sought to master *all* the magical and mysterious forces of nature.

He had a few followers during his lifetime, and many more after his death. They called themselves Paracelsians and continued to try to change medicine and science as he had done. They experimented in the laboratory and used chemical remedies in their medical practices. They tried, like Paracelsus, to control the forces of nature through natural magic.

The Paracelsians always remained outside the mainstream. The majority of doctors and scientists were unwilling to totally reject the legacy of the Ancients. Nevertheless, Paracelsus's message was increasingly picked up. People started looking at the world for themselves. In 1543, two years after his death, two books were published, one on anatomy, one on astronomy, which also challenged the authority of the Ancients. The universe was being looked at anew.

CHAPTER 10

Uncovering the Human Body

If you want to really understand how something is made, it is often a good idea to take it apart, piece by piece. With some things, like watches and cars, it helps if you also know how to put them back together again. If what you want to understand is a human or an animal body, it has to be dead before you start, but the goal is the same.

Galen, as we know, dissected – took apart – many animals, because he couldn't dissect any humans. He assumed that the anatomy of pigs or monkeys was pretty much like that of human beings, and in some ways he was right, but there are differences, too. The dissection of human bodies started to be done occasionally around 1300, when medical schools began to teach anatomy. At first, when people noticed any differences between what they saw in the human body and what Galen had said, they assumed that human beings had simply changed, not that Galen had been wrong! But as they began to look more closely, anatomists discovered more and more small differences. It became obvious that there was more to uncover about the human body.

The man who did the uncovering was an anatomist and surgeon known to us as Andreas Vesalius (1514–64). His full name was Andreas Wytinck van Wesel. He was born in Brussels, in modern-day Belgium, where his father was a medical man employed by the German Emperor Charles V. A clever child, he was sent to the University of Louvain to study arts subjects, but decided to change to medicine. Clearly ambitious, he then went to Paris where some of the best teachers were. They all followed Galen, and during his three years there he impressed them. He also showed his abilities in Greek and Latin, and his fascination with dissection. A war between the German Empire and France forced him to leave Paris, but he reintroduced human dissection to the medical faculty at Louvain before travelling, in 1537, to what at the time was the best medical school in the world, at the University of Padua in Italy. He took his exams, passed with the highest distinction, and the next day was appointed as a lecturer in surgery and anatomy. At Padua they knew when they were on to a good thing: Vesalius taught anatomy through his own dissections, the students loved him, and the very next year he published a series of beautiful anatomical illustrations of parts of the human body. They were so good that doctors all over Europe began copying these pictures for their own use, much to Vesalius's annoyance, since they were actually stealing his work.

Cutting open a dead body is not a particularly pleasant thing to do. After death, the body quickly begins to decay and smell and, in Vesalius's time, there was no way to stop it from rotting. This meant that the dissection had to be done quickly, and in an order that made it possible to get it done before the smells became over-powering. The belly was done first, since the intestines are the first to rot. This was followed by the head and brain, then the heart, lungs, and other organs in the chest cavity. The arms and legs were saved to the end: they lasted the best. The whole thing had to be done in two or three days, and anatomy was generally taught in winter, when the colder weather at least delayed the decay and gave the doctors a little more time.

Means of preserving bodies were discovered in the 1700s, and this made it easier to take more time to dissect and examine the whole body. When I was a medical student, I took eight months to dissect a body, and on dissection days my clothes and fingernails smelled not of the rotting body but of the preserving chemical. I worked on the body of an old man and I became very familiar with him during those months. The order we did things was pretty much the same as it was in Vesalius's time, except we saved the brain for last, since it is such a complicated organ and we were supposed to be better at carefully cutting out and exposing the different parts of the body by then. The old man had donated his body to science. He certainly taught me a lot.

Despite the speed needed, and the smells he confronted, dissection was Vesalius's great passion in life. We cannot know how many bodies he carefully cut apart, but it must have been many, for he came to know more about the parts of the human body than anyone then alive. The five and a half years between the time Vesalius became a teacher in Padua and the publication of his great book, in 1543, were very busy. Vesalius's book is enormous, forty centimetres high and weighing nearly two kilograms – not exactly a paperback you could slip in your pocket for holiday reading. It was called *De Humani coporis fabrica* ('On the structure of the human body'), and it is still known as *De Fabrica*. It was beautifully and intricately illustrated. Vesalius travelled to Basel, in Switzerland, to supervise the printing of the text and the making of the illustrations.

We live in a world where illustrations are everywhere. Digital cameras make it easy to send pictures to our friends, and magazines and newspapers have pictures on every page. It was not like that in Vesalius's day. The printing press had been invented less than a hundred years before, and pictures had to be made from carefully carved blocks of wood, copied from a drawing. Like a rubber stamp, these blocks were then inked and pressed on a piece of paper.

The pictures in Vesalius's book are staggering: never before had the human body been depicted so accurately, or in such detail. Even the title page tells us that something special is happening. It shows the dissection of a woman, in public, with hundreds of people crowding around. Vesalius stands in the middle, by the woman's body, and he is the only person looking out at the reader. The rest of the audience is either fascinated by the dissection or gossiping with each other. On the left of the picture is a monkey, on the right a dog, reminders that Galen had had to use animals for his anatomical work. In his own book, Vesalius is talking about human anatomy, from human bodies, and doing the dissecting himself. It was a wonderfully daring thing for a young man of not yet thirty.

But, then, Vesalius had every reason to be confident. He knew that he had seen further into the human body than anyone. Among the magnificent pictures in his book are those showing the muscles of the body, front and back, with the muscles nearer the surface dissected away to expose the deeper ones. These 'muscle men' are posed against landscapes, and the buildings, trees, rocks and hillsides in the pictures all join up. One of Vesalius's muscle men is being hanged by the neck, a reminder that Vesalius often used criminals for his dissections. Indeed, he once found a criminal who had been hanged and his body had been picked clean by birds, leaving only his skeleton. Vesalius smuggled the bones back to his room one by one, in order to study in private.

Vesalius had a very skilled artist to work with him, although we don't know his name for sure. Science was closely linked with art during this period, which we call the Renaissance, 're-birth'. Many Renaissance artists – Leonardo da Vinci (1452–1519), Michelangelo (1475–1564), and others – dissected bodies in order to learn how to paint them better. Doctors weren't the only ones who wanted to know about the structure of the human body.

Vesalius was fascinated by the structure (anatomy) of the body, but dead bodies do not carry out functions (physiology) like breathing, digesting and moving, as living ones do. So the long written part of Vesalius's book was a mix of old and new ideas. He often pointed out how Galen had described some organ or muscle incorrectly and he set it right. For instance, when Galen described the liver, he was talking about the pig's liver, which has five distinct 'lobes', or sections. The human liver has four, which are not so clearly defined. Several muscles in human hands and feet are different from those of even our close kin, monkeys and apes. Galen's theory of how the blood moves required a little of it to move from the right side of the heart to the left; he had it seeping through tiny pores in the wall between the two big chambers (ventricles) of the heart. Vesalius dissected many human hearts and could not find these pores. His knowledge would be very important a few decades later when William Harvey began to think in more detail about what the heart and blood do. Yet Vesalius's discussion of how the living body works still used many of Galen's ideas. This is perhaps why Vesalius's pictures were so much more valued than his writing: the pictures were soon copied and used throughout Europe, and made Vesalius famous (even if they did not earn him much money).

Although he lived for another twenty years, the publication of his great book was the highlight of Vesalius's career. He did produce a second edition of the book, with a few corrections, but soon after the first edition was published, he went off to be a court doctor. He spent his time taking care of the rich and powerful. Perhaps he thought he had said everything he had to say.

He had said and done enough to make sure that he was remembered. De Fabrica remains one of the great books of all time: a combination of art, anatomy and printing that is still admired today. And with it Vesalius left us two permanent gifts. First, he encouraged other doctors to continue his minute descriptions of the structure of the human body. Later anatomists discovered other parts of the body that Vesalius had missed, or corrected errors that he had made. The mix of artistic presentation and careful dissection that he had started encouraged others to produce books that illustrated the body on the page. Vesalius's book was the first in which the pictures were more important than the writing, but it was not the last. Doctors needed to be taught how to see what was before them, and pictures were essential to help them learn.

Second, Vesalius stood up to Galen. He wasn't rude about him, like Paracelsus, but he quietly showed that one could know more

than Galen had. He showed that knowledge can grow from generation to generation. He helped begin a debate that lasted for more
than a hundred years. The question was simple: Can we know
more than the Ancients? In the thousand years before Vesalius, the
answer had been No. After Vesalius, the answer began gradually to
change. People started to think: 'If everything worth knowing has
already been discovered, what's the point of bothering? But if I look
for myself, maybe I can see something that no one else has seen.'
Vesalius encouraged doctors and scientists to start bothering.

CHAPTER 11

Where is the Centre of the Universe?

Every morning, the sun rises in the east, and every evening, it sets in the west. We can see it slowly move throughout the day, with our shadows long or short, in front or behind, depending on where the sun is. Try the experiment at midday, and see your shadow tuck up under you. Nothing could be so obvious, and since it happens every day, if you miss it today you can catch the show tomorrow.

The sun doesn't go around the earth each day, of course. You can understand how difficult it would be to convince people that what seems so obvious is not really what is going on. Put it this way: the earth is the centre of *our* universe, because that is where we are when we look at the sun, moon and stars. It's our centre, but not *the* centre.

All the stargazers in the ancient world had put the earth at the centre. Remember Aristotle? After him, the most influential Greek astronomer, Ptolemy, built on the careful noting of the position of the stars night after night, season after season, and year after year.

Looking at the stars on a clear night is a magical experience, and being able to identify the groups, or 'constellations', of stars is great fun. The Plough and Orion's belt are easy to trace across the sky when there are no clouds. From the Plough you can find the North Star, and this helped sailors at night to continue to sail in the right direction.

There were problems with a model of the universe in which the earth is at the centre and the heavenly bodies move around it in perfect circles. Take the stars, for instance. They change their positions only gradually, as the nights pass. The spring equinox – when the sun is directly above the equator, making the day and night of equal length – has always been important for astronomers, and, in fact, for everyone. It occurs on either 20 or 21 March, and the 21st is the official first day of spring. The trouble is, the stars are in slightly different positions each first day of spring, which they shouldn't be if they were moving in perfect circles around the earth. Astronomers called this 'the precession of the equinoxes', and they had to make complicated calculations to explain why this happens.

The movement of the planets was also a puzzle. When you simply look at the night sky with your naked eyes, the planets appear as bright stars. Ancient astronomers thought that there were seven planets: Mercury, Venus, Mars, Jupiter and Saturn, plus the sun and the moon, which they also called planets. They were obviously closer to the earth than what they called the 'fixed stars', which we call the Milky Way. Observing the planets created more problems than the fixed stars, since they do not move as if they are circling the earth. For one thing, their movement does not appear to be constant, and the planets sometimes seem to go back on themselves. To solve this problem, astronomers said that the point that the planets were spinning around was not actually at the centre of the earth. They called this point the 'equant', and this and other calculations helped stargazers explain what they could see in the sky at night without having to throw away the model entirely. It meant that they could still assume that the earth was at the centre of things and that the other heavenly bodies revolved around it.

What would happen if instead of placing the earth at the centre of things, you put the sun there, and assumed that the planets (now including the earth as one of them) revolved around it? We are so accustomed to this view that it is hard to realise what a dramatic step it was. It went against what we see every day, it went against the teachings of Aristotle and (more importantly) of the Church, for in the Bible Joshua is said to have asked God to command the moving sun to stand still. But putting the sun at the centre of things was exactly what a Polish priest named Copernicus boldly did.

Nicholas Copernicus (1473–1543) was born and died in Poland, but he studied both law and medicine in Italy. His father died when Nicholas was ten years old, so his mother's brother took charge of educating the clever young boy, at the University of Cracow, in Poland. When his uncle became Bishop of Frauenburg, also in Poland, Copernicus obtained a job at the cathedral. This gave him a secure income, enabled him to study in Italy, and when he returned, to continue his passion: studying the heavens. He built a roofless tower, where he could use his astronomical instruments. Since there were not yet any telescopes, these instruments simply allowed him to measure the angles between various heavenly bodies and the horizon, and the phases of the moon. He was also very interested in eclipses, which occur when the sun, moon, or one of the planets gets in the way of another planet and becomes partly or wholly covered from our sight.

We don't know exactly when Copernicus decided that his model of the heavens and the solar system (as we now call it) was better at explaining the observations people had been making for thousands of years. But in 1514 he wrote a short manuscript and showed it to a few trusted friends. He did not dare publish it. In it, he stated quite clearly that 'the centre of the earth is not the centre of the universe', and 'we revolve around the sun like any other planet'. These were pretty definite conclusions, and during the next three decades, Copernicus quietly worked on his theory that the sun, not the earth, is at the centre of the universe. Although he spent many hours observing the heavens himself, he was at his best in thinking about what other astronomers had seen, and how their difficulties could be smoothed out by placing the sun at the centre and assuming that the planets rotated around it. Many puzzles, such as eclipses, or the strange forward and backward movement of the planets, fell into place. Besides, the sun has such an important role in human life, giving us warmth and light, that making it central was a way of recognising that without it, life on earth would be impossible.

Copernicus's model had another very significant consequence: it meant that the stars were much further away from the earth than Aristotle and other earlier thinkers had assumed. Aristotle thought that time was infinite but space was fixed. The Church had taught that time was fixed (to a few thousand years before, when God created everything), and so was space, except perhaps for Heaven itself. Copernicus accepted the Church's ideas of time and creation, but his measurements told him that the earth was much nearer to the sun than the sun was to the other stars. He also calculated the approximate distances from the sun to the planets, and of the moon from the earth. The universe was much larger than people had thought.

Copernicus knew his research would shock people, but as he got older, he decided that he should publish his ideas. In 1542, he finished his big book, *De revolutionibus orbium coelestium* ('Revolutions of the heavenly bodies'). But by then Copernicus was a sick old man. So he entrusted its printing to his friend, another priest called Rheticus, who knew about his ideas. Rheticus began the job, but then he had to go to work at a university in Germany, and the task was entrusted to yet another priest, Andreas Osiander. Osiander believed that Copernicus's ideas were dangerous, so he added his own introduction to this great book, which was finally printed in 1543. Here he wrote that Copernicus's ideas were not actually true, but were simply a possible way of solving some of the difficulties astronomers had long recognised with their earth-centred idea of the universe. Osiander was entitled to his own opinion, but he did a very dishonest thing: he wrote this preface as if it was the work of Copernicus himself. Since it was not signed by anybody, people assumed that this was what Copernicus meant to say about his ideas, and Copernicus was by then close to death and unable to do anything to correct the false impression that the preface gave. Consequently, for almost one hundred years, readers of this wonderful book assumed that Copernicus was merely playing around with ways to explain what you saw in the heavens each night, but not really saying that the earth went around the sun.

This preface made it easy for people to ignore the revolutionary message in Copernicus's book. Many people read it, however, and its comments and calculations influenced astronomy in the decades after he died. Two especially important astronomers took his work even further. One of them, Tycho Brahe (1546–1601), was inspired by Copernicus's insistence that the universe must be very large, so far away were the stars. Observing an eclipse of the sun in 1560 fired his imagination, and although his noble Danish family wanted him to study law, the only thing that really satisfied him was studying the heavens. In 1572, he noticed a new, very bright star in the night sky. He wrote about this *nova stella* ('new star') and argued that it showed that the heavens were not completely perfect and changeless. He built himself an elaborate observatory on an island off the coast of Denmark, and equipped it with the most advanced tools. (Alas, telescopes had still not been invented.) In 1577 he followed the path of a comet; these were generally seen as bad omens, but for Tycho, the comet's path merely signified that the heavenly bodies were not fixed in their own spheres, since the comet cut across them.

Tycho made many important discoveries about the positions and the movements of the stars and planets, although he eventually had to close his observatory and move to Prague, where in 1597 he established a new astronomical observatory. Three years later, he made Johannes Kepler (1571–1630) his assistant. Although Tycho never accepted Copernicus's model of the sun at the centre of things, Kepler had a different outlook on the universe, and Tycho left him all his notes and manuscripts when he died in 1601. Kepler was dutiful to Tycho's memory and edited some of his work for publication, but he also took astronomy into an entirely new direction.

Kepler had a stormy, chaotic life. His wife and young daughter died, and his mother was put on trial for witchcraft. He himself was an intensely religious Protestant in the early days of the Reformation, when most authorities were Catholic, so he had to watch his step. He believed that the order of the heavens confirmed his own mystical appreciation of God's creation. For all that, his lasting contribution to astronomy was very hard-nosed and precise. In the midst of his writings, which are often difficult to understand, he elaborated three concepts that are still known as Kepler's Laws. They were extremely important.

His first two laws were closely related, and his discovery of them was helped by the careful observations of the movements of the planet Mars that Tycho had left him. Kepler studied these for a long time before he realised that planets do not always move at the same speed; rather, they move faster when they are closer to the sun, and slower when they are further away from it. He found that if you draw a straight line from the sun (at the centre of the universe) to the planet, it is the curve of the arc made as the planet moves that is constant, not the planet's speed. This was his second law, and its consequence was his first law: that planets move not in perfect circles, but in ellipses, a kind of flattened circle. Although gravity had not yet been thought about, Kepler knew that some kind of force was acting upon the planets' movements. And he realised that the ellipse is the natural path of something revolving around a central point, as planets do around the sun. Kepler's two laws showed that the ancient idea of perfect circular motion in the heavens was wrong.

His third law was more practical: he showed that there is a special relationship between the time a planet takes to revolve completely around the sun and its average distance from the sun. This allowed astronomers to calculate the distances of the planets from the sun, and to get a sense of how large our solar system is, but also how small when compared to the enormous distances between us and the stars. Luckily, at around the same time a scientific instrument was invented to help us look further into those distances. The man who turned the telescope into a tool of immense power was the most famous astronomer of all: Galileo Galilei.

CHAPTER 12

Leaning Towers and Telescopes
GALILEO

One of the strangest buildings in the world must be the 850-year-old bell tower of the cathedral in the city of Pisa in Italy. You may know it as the Leaning Tower of Pisa. It's fun to take photographs of a friend in front of it pretending to hold the tilting tower from falling. There are also stories about how Galileo used the tower to perform his own experiments – dropping two balls of different weights from the top to see which would land first. In fact, Galileo didn't use the tower, but he did other experiments that showed him what the result would be, and he found that a ten-pound ball and a one-pound ball would hit the ground at the same time. Like the sun not moving around the earth each day, this experiment seemed to go against our everyday experience. After all, a feather and a ball dropped from the tower do not fall at the same rate. Why would the differently weighted balls drop to earth together?

Galileo Galilei (1564–1642) was born in Pisa. (Galilei was the family name, but our hero is always known by his first name.) His father was a musician and Galileo actually grew up in nearby Florence. He returned to the University of Pisa as a young man, starting to study medicine, but he was always more interested in mathematics, and he left the university with a reputation for cleverness and quick wit. In 1592, he went to Padua to teach mathematics and what we would call physics. He was there when William Harvey, whom we shall meet shortly, was a student, and it's a shame that the two probably never met each other.

Galileo attracted controversy throughout his life. His ideas always seemed to challenge accepted views, especially the physics and astronomy of Aristotle and the other Ancients. He was a good Catholic, but it was also his belief that religion is about morality and faith, whilst science deals with the observable, physical world. As he put it, the Bible teaches how to go to heaven, not how the heavens go. This brought him into conflict with the Catholic Church, which was energetically defending itself against those who dared to challenge either its ideas or its authority. The Church also started policing the growing number of books that were produced by the printing presses, placing unacceptable ones on a list they called the *Index Librorum Prohibitorum* – the 'List of Banned Books'. Galileo, who had many friends in high places (including princes, bishops, cardinals and even popes), had the support of many churchmen, but others were determined not to allow his ideas to upset their teachings, which were centuries-old.

Galileo's early work was with the forces involved in moving objects. From the very beginning he was someone who wanted to observe and measure things for himself, and if possible to express his results mathematically. In one of his most famous experiments, he carefully rolled a ball down a tilted surface and measured how long it took to reach certain distances. As you can imagine, the ball picks up speed as it moves down the slope (we would say it accelerates). Galileo saw that there was a special relationship between the speed of the ball and the time that had passed since it started moving. The speed was related to the square (a value multiplied by itself, such as 3 × 3) of the time taken. So, after two seconds, Galileo discovered that the ball would be travelling four times as fast. (The square of the time taken also appears in later scientists' work, so look out for it. Nature seems to like things squared.)

In all these, and many more experiments, Galileo showed himself to be a very modern scientist, because he knew that his actual measurements were not always exactly the same; sometimes we blink at a bad moment, or it takes time for us to record what we see, or the equipment isn't perfect. However, these are the kinds of observations we can make about the real world, and Galileo was always most interested in the world as we find it, not in some abstract world where everything was always perfect and exact.

Galileo's early work on moving objects showed how differently he saw the world as compared with Aristotle and the hundreds of thinkers who had come after, despite Aristotle's continuing importance in the universities, which were governed by religious groups. In 1609, Galileo learned of a new instrument that would challenge the ancient way of thinking even more seriously. This instrument was soon to be called the 'telescope', a word that means 'to see far', just as 'telephone' means 'to speak far', and 'microscope' means 'to see small'. Both telescopes and microscopes have been very important in the history of science.

The first telescope that Galileo constructed offered only a little magnification, but he was very impressed with it. He quickly improved it by combining two lenses, so that he could get the kind of magnifying power that we expect from an ordinary pair of binoculars today, about fifteen times. That doesn't sound like much, but it created a sensation. Using it, one could spot ships coming in from the sea long before they were visible to the naked eye. More importantly, Galileo turned his telescope to the heavens and was amazed at what he found there.

When he looked at the moon, he realised that it was not the perfect, smooth, circular ball that people had supposed. It had mountains and craters. Turning his telescope towards the planets, he observed their movements more closely, and discovered that one planet, Jupiter, had 'moons' just as the earth had its moon. Another planet, Saturn, had two big blobs which didn't look like moons and which we now call its 'rings'. He could see the movements of Venus and Mars much more clearly, and agreed that they changed their direction and speed in a regular and predictable way. The sun had dark areas or spots, which moved a bit each day in regular patterns. (He learned to look at it indirectly, to protect his eyes, as you must.) His telescope revealed that the Milky Way, which appears as a wonderful, fuzzy blur of light when looking with the naked eye on a clear night, was actually composed of thousands and thousands of individual stars, very far away from the earth.

With his telescope, Galileo made these and many other important observations. He wrote about them in a book called *Starry Messenger* (1610), which created a stir. Each revelation called into question what people thought about the heavens. Some thought Galileo's ideas were based on tricks played by his new 'tube', as the telescope was often called, because what could not be seen by the naked eye might not be there. Galileo had to try to convince people that what his telescope showed was real.

Much more awkwardly, and dangerously, was that Galileo's observations were good evidence for Copernicus being right about the moon revolving around the earth, and about the earth, moon and the other planets all orbiting around the sun. By this time, Copernicus's book had been in print for almost seventy years, and he had a number of followers, Protestants as well as Catholics. The official position of the Catholic Church was that Copernicus's ideas were useful to work out the movements of the planets, but they were not literally true. If they were, too many passages from the Bible would be complicated, and have to be thought about again.

But Galileo wanted to tell people about his astronomical findings. He went to Rome in 1615 hoping to get the Church's permission to teach what he had learned. Many people – even the Pope – sympathised with him, but he was still forbidden to write about, or teach, Copernicus's system. He didn't give up entirely, going to Rome again in 1624 and 1630 to test the waters, though he was getting old and unwell. He became convinced that as long as he was careful to present the Copernican system only as a possibility then he would be safe. His work on astronomy, *Dialogue on the Two Chief World Systems*, is written as a conversation between three people: one representing Aristotle, another representing Copernicus, and the third acting as the host. That way, Galileo could discuss the pros and cons of old and new ideas about the universe without having to say which was right or wrong.

It is a wonderful book, full of jokes, and written, like most of Galileo's works, in his native language, Italian. (Scholars from all over Europe still usually wrote their books in Latin.) From the start, it was pretty obvious which side Galileo was on. For one thing, the Aristotelian character was named Simplicio. Now, there was in fact an ancient commentator on Aristotle called that, but just as in English, in Italian it sounds like 'simpleton', and this character isn't very bright. The Copernican (called Salviati, a name that suggests 'wise' and 'safe') has by far the best lines and arguments.

Galileo tried very hard to get the Church's official approval for his book. The censor in Rome, who controlled which books could be published, was sympathetic to Galileo, but he knew there would be problems and so delayed his decision. Galileo went ahead and had the book printed in Florence. When the high churchmen in Rome read it, they were not pleased, and summoned the old man to Rome. Someone dug out a copy of the old ban against him teaching the Copernican system, and after a 'trial' in 1633 that went on for three months, Galileo was forced to say his book was an error and the product of his vanity. The earth, he said in his signed confession, does not move and is the centre of the universe. There is a legend that immediately after being convicted, Galileo muttered, 'Eppur si muove' ('And yet it moves'). Whether or not he did say it out loud, he certainly thought it, for the Church could not force him to change his beliefs about the nature of the world.

The Church had the power to throw Galileo in prison and even torture him, but his jury recognised that he was a very unusual man, and put him under house arrest instead. His first 'house arrest' in the city of Siena wasn't all that strict – he was the life and soul of many dinner parties – so the Church insisted that he return to his home outside Florence, where his visitors were carefully policed. One of Galileo's daughters (a nun) died soon after, and his last years were lonely. But he continued his work, returning to the problems of falling objects and the forces that produce the kinds of movement we see around us every day. His great work, *Two New Sciences* (1638), is one of the foundations of modern physics. He looked again at the acceleration of falling bodies, and used mathematics to show that acceleration could be measured in a way that anticipated Isaac Newton's later famous work on gravity. He also offered a new way of thinking about the paths of things shot through the air, like cannon-balls, showing how it could be predicted where they would land. With this work, the concept of 'force' – what influences something to move in a particular way – took its place in the study of physics.

If you've ever heard the phrase 'rebel without a cause', then Galileo was a rebel *with* a cause. The thing he fought for was science as knowledge that can explain the way the world works in its own terms. Some of his rebellious ideas were later abandoned because they were wrong, or did not fully explain things. But that's the way science always works, and no area of science is a closed book containing all the answers. Just as all modern scientists should, Galileo knew this.

CHAPTER 13

Round and Round
HARVEY

The words 'cycle' and 'circulation' are both based on the original Latin word for 'circle'. Going through a cycle, or circulating, means you just keep moving and eventually come back to where you started from, without necessarily noticing you are back at the beginning. There are not many perfect circles in nature, but there is a lot of circulation. The earth circles around the sun. Water circulates by evaporating from the earth and falling again as rain. Many birds migrate long distances each year, then return to the same area to breed and start their yearly cycle over again. Indeed, the whole natural process of birth, growth and death, followed by the repeat of the cycle in a new generation, is a kind of circulation.

There are also a lot of cycles, or circulations, within our bodies. One of the most important of these involves the heart and blood. Each drop of blood circulates through our bodies about fifty times every hour of our lives. That varies, of course, depending on what we're doing: if we're running, and our hearts have to beat faster, the circulation time is shortened; when we're asleep, our hearts beat more slowly and it takes longer for a drop of blood to get back to the heart. These days, we learn all this in school, but it was not always so clear-cut. The man who discovered that our blood circulates was an English doctor named William Harvey (1578–1657).

Harvey's father was a farmer who became a successful merchant, an occupation that five of Harvey's six brothers followed too. William Harvey chose medicine as a career, however, and after finishing his medical studies at Cambridge University in 1600, he went to the University of Padua, where Vesalius had worked a few years before, and where Galileo was currently investigating astronomy and physics.

One of Harvey's medical teachers at Padua was Fabrizi of Acquapendente (1537–1619). Fabrizi was continuing the research tradition started long before by Aristotle, and it inspired Harvey. Teacher and pupil absorbed two important lessons from Aristotle. First, that in living creatures, including human beings, the organs in our bodies have the form, or structure, they have because of the work they have to do. Our bones and muscles, for instance, are put together so that we can run, or pick up things, and unless there is something wrong with us we don't even notice them functioning in the way they seemed designed for. Aristotle also believed that everything within plants and animals had a specific purpose, or function, because the Creator wouldn't design any parts that were useless. Our eyes are constructed the way they are so that we can see; so are the other parts of our bodies, our stomach, liver, lungs and heart. Each organ has a special structure, in order to perform its own particular function. This approach to understanding the way our bodies work was called 'living anatomy', and it was especially helpful in figuring out the 'logic' of how our bodies operate. It was clear to doctors that bones were hard, and kept their shape, because they have to support our bodies when we are walking or running. Our muscles are softer and springier because their contraction and relaxation helps us move. Yet it was not so obvious that the heart, and its relationship to the blood and blood vessels, could be understood using the same logic. Perhaps we should say that the heart now fits into this way of thinking about our bodily functions because we have Harvey to guide us.

Second, Aristotle insisted on the central role that the heart and blood play in our lives, after observing the tiny beating heart which was the first sign of life in the speck of a chick in an egg. Aristotle convinced Harvey that the heart is at the centre of life. And the heart and circulation became the centre of Harvey's medical career.

Harvey's own teacher, Fabrizi, also discovered something that became crucial to Harvey: that many of the larger veins have valves in them. These valves are always situated so that the blood can go only one way: towards the heart. Fabrizi thought that their function was to prevent the blood pooling in our legs, or from rushing down from the brain with too great a force. Harvey made use of all these lessons when he returned to England after he completed his studies at Padua.

Harvey's career went from strength to strength. He set up a medical practice in London, got a job at St Bartholomew's Hospital, and was also soon being asked to lecture to surgeons on anatomy and physiology. He became a doctor to two kings of England, James I and then his son Charles I. Being associated with Charles I didn't help Harvey during this period, especially after the king was removed from the throne by a group of Protestants called Puritans. On one occasion, Harvey's house was attacked and burned, and with it many manuscripts for books he hoped to publish. This was a great loss to science, since Harvey had been investigating many things including breathing, muscles, and how animals form from fertilised eggs. King Charles had even allowed some of his own royal animals to be used in Harvey's experiments.

Harvey was always fascinated by blood. He thought it was really the essential part of what it means to be alive. He too cracked open some eggs and saw that the first sign of life was a speck of blood, pulsing in a rhythmical way. The same was true for other animals he examined when they were still embryos (still developing in the egg or their mother's womb). The heart, which has long been associated with blood, was also fascinating to Harvey. Everyone knew that when the heart stopped beating, the person or animal died. So, while blood was essential to the beginning of life, life ended when the heart stopped beating.

Most of the time our heart beats without us thinking twice about it. But sometimes you can actually feel your heart beating, for example, when you are nervous or scared, or when you have been exercising, and you feel your heart pummelling against your chest wall: duh-dum, duh-dum, duh-dum. Harvey wanted to understand the 'motions' of the heart, that is, what actually happens in each heartbeat. In every beat of the heart, the heart contracts (a process known as the 'systole') and then relaxes (the 'diastole'). He dissected many live animals in order to observe their beating hearts, especially snakes and other cold-blooded animals (those which can't regulate their own body temperatures). Their hearts beat much slower than ours do, so he could see the beating more easily. He saw how the valves inside the heart open and close, in every heartbeat, in a regular sequence of events. During contraction, the valves between the chambers of the heart closed, and those that connected the heart to the blood vessels opened. As the heart relaxed, the reverse happened, and the internal valves opened, while those that sat between the heart and the blood vessels (the pulmonary artery and the aorta) shut. It occurred to Harvey that these valves act just like the valves of the veins that his teacher Fabrizi had discovered, and that their function seemed to keep the blood going in a constant direction.

Harvey did several experiments to help others see what he was thinking. One was very simple. He placed a tight bandage (called a tourniquet) around an arm: if it was very tight, so no blood could get into the arm at all, the hand became very pale; if he loosened it a bit, the blood could get in but could not get back to the heart, and the hand became very red. This showed that the blood entered the arm at a certain pressure, which the tight tourniquet blocked entirely. Loosening the strap allowed blood to come in through the arteries, but not to get back out of the arm through the veins.

Having looked at so many hearts and thought so deeply about them, Harvey made an important leap in our understanding of what they do. He worked out that in a very short space of time more blood than was contained in the entire body passes through the heart. And it was impossible to make enough blood for each new heartbeat to pump new blood, let alone for a human body to contain it all. Therefore, the blood must go from the heart with each beat, travel through the arteries, into the veins, and return to the heart to begin a new cycle of 'circulation'.

'I began privately to consider that the blood had a movement, as it were, in a circle.' He wrote these words (in Latin) in 1628, in a short book called *De motu cordis* ('On the motion of the heart'). It seems as if he started out to write something on the contraction and relaxation of the heart, and ended up discovering what function these processes perform. He worked out that blood is pumped into the lungs (from the heart's right chamber), and also into the biggest artery, the aorta, from the left. From the aorta, the blood goes into the smaller arteries that branch off it, and then transfers to the veins, where the valves ensure that it flows in the correct direction and is returned to the right side of the heart through the largest vein, the *vena cava*.

Like Vesalius, Harvey always insisted that he wished to learn about the structures and functions of the body from his own investigations, not simply from books written by others. Unlike Vesalius, he worked mostly with living animals, not human corpses. He did not set out to challenge 2,000 years of medical teaching about the heart and blood, but he knew his findings would be controversial, because they showed that Galen's theory of the heart and blood was wrong. He defended his ideas against criticism from some people, mostly followers of Galen, who thought that his ideas were too extreme. But there was one important gap in his theory: he could not answer the crucial question of how the blood gets from the smallest arteries to the smallest veins, to begin its return journey back to the heart.

That bit of the puzzle was solved about the time of Harvey's death by one of his Italian disciples, Marcello Malpighi (1628–94), who was an expert at using a new instrument called the microscope, which had been around since the 1590s but was improved by Malpighi's time. He was able to look more closely than anyone before at the delicate structures of the lung, the kidneys and other organs, and he uncovered the tiny channels connecting the smallest arteries and veins: the capillaries. Harvey's 'circle' was complete.

Through his ground-breaking work, Harvey had shown what careful experimentation could uncover, and as his ideas became more widely accepted, people recognised him as a founder of experimentation in biology and medicine. This encouraged others to look for themselves and investigate other bodily functions such as what happens in the lungs when we breathe, or in the stomach when we digest our food. And, like Vesalius and Galileo before him, he helped people realise that scientific knowledge can increase, and that we can know more about nature than equally clever people who lived a thousand (or even fifty) years before us.

CHAPTER 14

Knowledge is Power
BACON AND DESCARTES

In the century between Copernicus and Galileo, science had turned the world upside down. The earth was no longer at the centre of the universe, and new discoveries in anatomy, physiology, chemistry and physics reminded people that the Ancients did not know everything after all. There was a lot out there still to be discovered.

People also started thinking about science itself. What was the best way to do it? How could we be sure that new discoveries were accurate? And how could we use science to improve our comfort, health and happiness? Two individuals in particular thought deeply about science: one an English lawyer and politician, the other a French philosopher.

The Englishman was Francis Bacon (1561–1626). His father, Nicholas Bacon, rose from humble beginnings to become a powerful official for Queen Elizabeth I. Nicholas knew how important education is, so he sent his son to the University of Cambridge.

Francis, too, served Elizabeth, as well as King James I, after Elizabeth died. He was an expert on English law, took part in several important trials and, after he became Lord Chancellor, was one of the major legal figures of his time. He was also active as a member of parliament.

Bacon was very enthusiastic about science. He spent a lot of time doing chemistry experiments and observing all kinds of curious things in nature, from plants and animals to weather and magnetism. More important than any discovery he made were his elegant and persuasive arguments about why science was worth doing, and how it should be done. Bacon urged people to value science. 'Knowledge is power,' he famously said, and science is the best way to achieve that knowledge. So he encouraged Elizabeth and James to use public money to build laboratories and provide places for scientists to do their work. Scientists, he thought, should form societies, or academies, so they could meet and exchange their ideas and observations. Science, he said, offers humans the means to understand nature, and, by understanding, to be able to control her.

Bacon wrote clearly about the best way for science to advance. Scientists needed to make sure that the words they used were precise and easily understood by others. They needed to approach their investigations with open minds, instead of trying to prove what they thought they already knew. Above all, they must repeat their experiments and observations, so that they can be certain of their results. This is the method of *induction*. For example, by counting, weighing or mixing chemicals again and again, the chemist can become properly confident about what is going on. As scientists collect more and more observations, or inductions, they will become surer about what will happen. They can use these inductions to form generalisations, which in turn will show them the laws governing how nature works. Bacon's ideas continued to inspire scientists for many generations. They still do so today.

So, in different ways, did those of the Frenchman René Descartes (1596–1650). He thought deeply about the work of both Harvey

and Galileo. Like Galileo, Descartes was a Catholic who nevertheless passionately believed that religion should not come into the
study of the natural world. Like Harvey, Descartes examined
human and animal bodies and explained how they worked in ways
that went far beyond what Galen had taught. In fact, even more
than either Harvey or Galileo, Descartes tried to establish both
science and philosophy on entirely new foundations. Although
today we remember him mostly as a philosopher, he was much
more of a practising scientist than Bacon.

Descartes was born in La Haye, in Touraine, France. A clever
boy, he went to a famous school, La Flèche, in the Loire region,
where excellent French wines are made. At La Flèche, he learned of
Galileo's discoveries with his telescope, Copernicus's placing the
sun at the centre of the universe, and the latest mathematics. He
graduated in law at the University of Poitiers, and then he did a
very surprising thing: he volunteered for an army of Protestants.
War raged in Europe during the whole of Descartes' adult life (the
Thirty Years War), and for almost nine years, he was part of it.
Descartes never actually fought, although his knowledge of practical mathematics, and where cannon-balls might land, could have
helped the soldiers. He was attached to both Protestant and
Catholic armies during these years, and seemed always to be where
important political or military events were taking place. We don't
know what he was doing, or how he got the money to travel so
much. Perhaps he was a spy. If so, it was probably for the Catholics,
to whom he always remained loyal.

Early in his adventures, on 10 November 1619, in a warm, stove-
lit room, half asleep, half awake, he came to two conclusions. First,
if he were ever to come to true knowledge, he had to do it all
himself. The teachings of Aristotle and other authorities would not
do. He needed to start over. Second, he concluded that the only
way to start over was simply by doubting everything! Later that
night, he had three dreams that he understood as encouraging this
idea. He didn't publish anything then, and in any case, his military
adventures had just begun. But this decisive day (and night) started
him on his path to explain the universe and everything in it, as well

as the rules that might help others obtain scientific knowledge with confidence.

Doubting everything meant taking nothing for granted, and then, bit by bit, following your nose by accepting only things you can be sure about. But what could he be sure about? In the first instance, only one thing: that *he* was planning this scientific and philosophical project. He was thinking about how to arrive at certain knowledge, but, more simply, he was *thinking*. 'Cogito, ergo sum,' he wrote in Latin: 'I think, therefore, I am.' I exist because I am thinking these thoughts.

This simple statement was Descartes' starting-point. That is all well and good, we might say, but what next? For Descartes, it had one immediate and far-reaching consequence: I exist because I am thinking, but I can imagine that I could think without having a body. However, if I had a body and couldn't think, I wouldn't know it. Therefore, my body and the thinking part (my mind, or soul) must be separate and distinct. This was the basis of *dualism*, the notion that the universe is made up of two completely different kinds of things: *matter* (for instance, human bodies, but also chairs, stones, planets, cats and dogs) and *spirit* (the human soul or mind). Descartes thus insisted that our minds – how we know we exist – have a very special place in the universe.

Now, people before and long after Descartes recognised that human beings are a special kind of animal. We have the ability to do things that no other animal has: to read and write, to make sense of the complexities of the world, to build jet planes and atomic bombs. Specialness was not the unusual part of Descartes' separation of our minds and our bodies. The amazing step was what he did with the rest of the world, the material part. Mind and matter are what the world is made up of, he said, and matter is the subject of science. This means that the material, non-thinking, parts of how we function can be understood in simple physical terms. And it means that all plants and all other animals, none of which have a soul, can also be completely reduced to matter doing its stuff. Along with trees and flowers, the fish and elephants are nothing but more or less complicated machines.

According to Descartes, they are things that can be completely understood.

Descartes knew about *automata*, mechanical lifelike figures specially made to move and do certain things. We would call them robots. For example, a lot of seventeenth-century town clocks had little mechanical figures, often a man coming out on the hour to strike a gong. They were all the rage in Descartes' day (and some still work now). People had already wondered if – since human beings could make such delicate figures, able to move and imitate humans or animals – perhaps a better mechanic could go one step further and make a dog that could eat and bark, as well as move. Descartes had no desire to make these toys, but in his thinking, plants and animals were just extremely complicated automata, with no real feelings and only the capacity to respond to what was happening around them. These machines were matter, which could be understood by scientists in terms of mechanical and chemical principles. Descartes read William Harvey's work on the 'mechanical' actions of the heart and the circulation of the blood, and he believed that this provided evidence for his system. (His own explanation of what goes on when the blood reaches the heart, and why it circulates, has been forgotten.) Descartes had great hopes that such ideas could explain much about health and disease, and ultimately offer human beings the knowledge of how to live, if not forever, at least for a very long time.

Having shown to his satisfaction that the universe is composed of two separate kinds of thing, matter and mind, Descartes puzzled how the human mind and its body were actually connected. He asked himself how they *could* be connected, if matter has substance and occupies space, and mind is the opposite, located nowhere and without any material basis at all. It had been common since the time of Hippocrates to associate our thinking powers with the brain. A blow to the head could knock a person out, and many medical men had observed that injuries and diseases of the brain led to changes in our mental functions. At one point, Descartes seemed to think that the human soul is located in a gland, in the middle of our brains, but he knew that, according to the logic of the system he had created, matter and mind could never physically interact. People later called this model of human beings 'the ghost in the machine', meaning that our machine-like bodies were somehow driven by a ghost-like mind, or soul. The problem then was to explain how many dogs, chimpanzees, horses and other animals share so many of our mental capacities without having their own 'ghosts'. Dogs and cats can show fear or anger, and dogs at least seem to be able to express love for their owners. (Cats are a law unto themselves.)

Descartes' curious mind puzzled over many other things: not surprising for someone who wrote a book called simply *Le Monde* ('The World'). He accepted Copernicus's ideas about the relationship between the earth and the sun, but was more careful than Galileo had been in presenting his ideas so that he did not offend the Church authorities. He also wrote about motion, falling objects, and other problems that attracted Galileo. Unfortunately, despite having some followers in his day, Descartes' ideas about how the universe works could not compete with those of giants like Galileo and Isaac Newton, and few remember Descartes' physics today.

If he lost out to clever men in the physics class, whether you know it or not, you follow in Descartes' footsteps every time you solve problems in algebra and geometry. Descartes had the bright idea of using a, b, c in algebra problems to stand for the known, and x, y, z to stand for the unknown. So when you are asked to solve an equation such as $x = a + b^2$, you are continuing the practice that Descartes started. And when you plot something on a graph, with a horizontal and a vertical axis, you are also using his invention. Descartes himself solved various algebraic and geometric problems in his book on those subjects, published along with the one on the world.

By so sharply separating body and mind, the material and the mental worlds, Descartes stressed how important the material world is for science. Astronomy, physics and chemistry deal with matter. So does biology, and if his idea of the animal-machine seems a bit far-fetched, biologists and doctors still try to understand how plants and animals function in terms of their material parts. It was just unfortunate for Descartes that his idea that medicine would quickly show people how to live for much longer was a bit before its time. He himself was pretty healthy until he accepted an invitation to go to Sweden to teach the Swedish queen his philosophy and knowledge of the world. She rose early and insisted that he give her the lessons very early in the morning. Descartes hated the cold. He did not survive even his first winter in Sweden. Catching some kind of infection, he died in February 1650, seven weeks before his fifty-fourth birthday. It was a sad end for someone who believed that he would live at least a hundred years.

Bacon and Descartes had lofty ideals for science. They differed in their ideas about how science could advance, but were passionate that it should. Bacon's vision was of science as a shared, state-funded enterprise. Descartes was more content to work things out by himself. Both wanted other people to take on and develop their ideas. Both men also believed that science is a special activity, superior to the humdrum of ordinary life. It deserved to be singled out in this way because science adds to our stock of knowledge and our ability to understand nature. Such understanding could improve our lives and the public good.

CHAPTER 15

The 'New Chemistry'

If you have a chemistry set then you may already know about litmus paper. These small strips of special paper can tell you whether a solution is acid or alkaline. If you stir some vinegar in water (making it acidic) and dip in the blue paper, it will turn red. If you try it with bleach (which is alkaline), the red paper will turn blue. Next time you use a piece of litmus paper, think of Robert Boyle, for he created the test more than 300 years ago.

Boyle (1627–91) was born into a large aristocratic family in Ireland. He was the youngest son, and never had to worry about money. Unlike a lot of wealthy people, Boyle was always generous with his fortune, and he donated a good deal of it to charity. He paid for the Bible to be translated into an American Indian language. Religion and science played equally big parts in his life.

He spent a few years at Eton, the elite English school, and then travelled in Europe, where he had a series of private tutors. Boyle returned to England where the Civil War was raging; some of his family sided with King Charles I, and some with the Parliamentarians, who sought to overthrow the king and establish a republic. His sister convinced him to join the Parliamentarians, and through her, he met an enthusiastic social, political and scientific reformer called Samuel Hartlib. Like Francis Bacon, Hartlib believed that science had the power to improve the lives of human beings, and convinced the young Boyle that studying agriculture and medicine could lead to such improvements. Boyle began with medicine and looked at the cures for various diseases, gaining along the way a lifelong fascination with chemistry.

Some religious people fear exposing themselves or their children to new ideas because they think the ideas might undermine their faith. Robert Boyle was not one of these people: his religious belief was so secure that he read whatever was related to his wide scientific interests. Descartes and Galileo were controversial figures in Boyle's early days, but he studied them both carefully – he read Galileo's *Starry Messenger* in 1642 in Florence, the very same year and place in which Galileo died – and used their insights in his own work. Boyle was also interested in the atomists of ancient times (Chapter 3), though he was not altogether convinced by their belief that the universe consists of nothing but 'atoms and the void'. He knew, however, that there were some basic units of matter in the universe, which he called 'corpuscles', but he could go about his work without the godless (atheistic) associations of ancient Greek atomism.

Boyle was equally unsatisfied with Aristotle's theory of the four elements – air, earth, fire and water – and he showed by experiment that it was not correct. He burnt a stick of fresh wood and showed that the smoke that came off it was not air. Nor was the fluid that oozed out of the end of the burning wood ordinary water. The flame differed depending on what was burnt, so that was not pure fire, and the ash that was left was not earth. By carefully analysing the results of these simple experiments, Boyle did enough to show that something as common as wood was not made of air, earth, fire and water. He also pointed out that some substances, like gold, could not be broken down further. When heated, gold melted and ran but it didn't change like wood did when it was burnt: when gold grew cold, it returned to its original form. Boyle recognised that the things that surround us in our daily lives, such as wooden tables and chairs, and woollen dresses and hats, were made up of a variety of components, but they could not be reduced to the four Greek elements, or to the three elements of Paracelsus. Some believe that Boyle came up with the modern definition of a chemical element. He certainly came close when he described elements as things 'not being made of other bodies, or of one another'. But he didn't take this any further, nor did he use it in his own chemical experiments.

Instead, Boyle's notion of the 'corpuscle' as a unit of matter suited his experimental purposes very well. Boyle was a tireless experimenter, spending hours in his private laboratory either alone or with friends, and writing up his experiments in great detail in books. It is partly this attention to detail that makes Boyle so special in the history of science. He and his friends wanted science to be open and public, and for others to be able to use the knowledge they gained. No longer was it enough to claim to have found out some deep secret of nature, as Paracelsus had done. A scientist needed to be able to demonstrate that deep secret to others, either in person or through written descriptions.

This insistence on openness was one of the guiding rules in the scientific circles in which Boyle moved. The first of these was an informal group in Oxford, where he lived in the 1650s; when most of the group moved to London, they joined with others to establish what became, in 1662, the Royal Society of London, still one of the leading scientific societies in the world. They knew that they were doing something that Francis Bacon had called for half a century earlier. Boyle was a leading light in this club devoted to increasing knowledge. From the beginning, the Fellows – as the Royal Society's members were called – were keen that the new knowledge they uncovered and discussed at their meetings should be useful.

One of Boyle's favourite collaborators was another Robert, a few years younger than him: Robert Hooke (1635–1702). Hooke was even cleverer than Boyle, but unlike Boyle, he came from a poor family. He always had to earn his way in life by his wits. Hooke was employed by the Royal Society to perform experiments at each of its meetings. He became very skilled at inventing and handling all kinds of scientific equipment. Hooke devised many experiments; for example, to measure the speed of sound, or to examine what happens when blood is transfused from one dog to another. In some cases the dog that had been given new blood seemed more energetic, and the men were encouraged to experiment with humans. They transfused blood from a lamb into a human being, but it didn't work; in Paris, too, one person who had been given a transfusion died, so these experiments were given up. Hooke's task at the Royal Society's weekly meetings was to prepare two or three less deadly experiments to entertain and stimulate the Fellows.

Hooke was one of the earliest 'savants' to make good use of the microscope. (A 'savant' literally means 'one who knows', and the term was often used to describe what we would now call scientists.) He used his microscope to reveal a new world of things invisible to the naked eye, uncovering structures in plants, animals and other objects that could never be seen without using it. The fellows loved to peer through the microscope at their meetings, and in addition to Hooke's demonstrations, they also received many communications from another famous early microscopist, a Dutchman named Antonie van Leeuwenhoek (1632–1723). Leeuwenhoek worked as a cloth merchant, but in his spare time he ground and polished very small lenses that could magnify things more than 200 times. He had to make a new lens for each observation, and crafted hundreds during his long life. He would place each lens in a metal holder with the small object that he wanted to examine behind it. He found tiny organisms in pond water, bacteria in the scrapings of his teeth and many other wondrous things. Hooke too believed that his microscope could take the observer closer to nature, and the illustrations in his book, *Micrographia*, published in 1665 (the very year of the London plague), caused a sensation. Many of these illustrations look odd to us, for they show very large, magnified insects, such as flies or lice, and these pictures have become quite famous. Yet he also filled his book with observations and speculations on the structures and functions of other things he could see through his microscope. He showed one picture of a thin section of cork, from the cork tree – the material used to close wine bottles. He called the little boxy structures he saw there 'cells'. They weren't actually what we now call cells, but the name stuck.

Both Boyle and Hooke had a favourite mechanical device: their version of the air pump. Hooke and Boyle's air pump worked in the same way as the pumps we use to put air into bicycle tyres or footballs. It had a large central cavity, with a tight fitting that could be opened at the top, and another opening in the bottom, where there was a valve through which gases could be drawn in or let out. It might not seem very exciting, but it helped solve one of the major puzzles of science during the period: whether it was possible to have a vacuum, that is, completely empty space, not even containing air. Descartes had insisted that vacuums were impossible ('Nature abhors a vacuum' was the common phrase expressing this idea). But if, as Boyle had argued, matter was ultimately composed of separate corpuscles, in different forms, there ought to be some space between them. If something like water is heated, so that it evaporates and turns into a gas, the same corpuscles would still be there, said Boyle, but the gas occupies more space than the liquid had done. After lots of experiments heating liquids to gases, he saw that all gases behaved pretty much the same when they were in the air pump. Boyle and Hooke came to a conclusion that is still known as Boyle's Law. At a constant temperature, the volume that any gas occupies has a special mathematical relationship to the pressure that it is under. We say that its volume is directly influenced by the pressure around it. So, if you increase the pressure by decreasing the space it occupies, the gas squeezes into the available space. (If you increase the temperature, the gas expands, and a new pressure comes into effect, but it's the same basic principle.) In the future, Boyle's Law would help the development of the steam engine, so remember him when we get there.

Boyle and Hooke used their air pump to examine the characteristics of many gases, including the 'air' that we breathe. Air was, remember, one of the Ancients' elements, but it was becoming clear to many people in the seventeenth century that the air that surrounds us and keeps us alive is not a simple substance. It was obviously involved in breathing, since we draw air into our lungs when we take a breath. But what else did it do? Boyle and Hooke, both individually and together, were very interested in what happens when a piece of wood or charcoal burns. They also wondered why blood was dark red before it went into the lungs and bright red when it came out of them. Hooke linked these two questions together and suggested that what happens in the lungs is a special kind of combustion, with the 'air' being the substance that connected both the breathing and the burning. Hooke pretty much left it at that, but the problems surrounding both the composition and nature of 'air', as well as what happens during respiration (breathing) and combustion, continued to intrigue scientists for more than a century after Boyle and Hooke, as people repeated and developed their experiments.

There was hardly any area of science that Robert Hooke did not think about. He invented a watch run by a set of springs (a great improvement in time-keeping), wondered about the origin of fossils, and investigated the nature of light. He also had brilliant things to say about a problem we have encountered before, and will look at in more detail in the next chapter: the physics of movement and force. Hooke was investigating these subjects at the same time as Isaac Newton. As we shall see, Newton himself is one of the reasons that everybody has heard of Sir Isaac, but few people know about Mr Hooke.

CHAPTER 16

What Goes Up . . .
NEWTON

I doubt if you have ever met anyone as smart as Isaac Newton –
I haven't. You might have met people as unpleasant as he was.
He disliked most people, had temper tantrums, and thought that
almost everybody was out to get him. He was secretive, vain and
would forget to eat his meals. He had lots of other disagreeable
characteristics, but he *was* clever, and it's the cleverness that we
remember today, even if it's quite hard to understand what he
thought and wrote.

Isaac Newton (1642–1727) might have been disagreeable no
matter what had happened to him, but his childhood was pretty
awful. His father died before he was born, and his mother, who
didn't expect him to live, left him with her parents after she
remarried and had another family. He hated his stepfather, disliked
his grandfather and wasn't very fond of either his mother or his
grandmother. In fact, from an early age, he started not liking
people. He preferred to be alone, as a child and as a very old man.

It was obvious, however, that he was very smart, and he was sent to the grammar school in Grantham, near where he lived, in Lincolnshire. He learned good Latin (he could write in English and Latin with equal ease), but spent most of his time at school making models of clocks and other mechanical gadgets and constructing sundials.

He also did his own thing when he went up to Trinity College, Cambridge, in 1661. He was supposed to read the ancient masters such as Aristotle and Plato. He did read them a little (he was a meticulous note-taker, so we know what he read), but his favourites were the moderns: Descartes, Boyle and other exponents of the new science. Reading was all right, but he wanted to figure things out for himself. To do this he devised many new experiments, but his greatest genius was in mathematics and how it could be used to understand more about the universe.

Newton worked out many of his ideas in an incredibly productive couple of years. No scientist except Einstein (Chapter 32) has ever done so much in so short a space of time. Newton's most amazing years were 1665 and 1666. Some of this time he spent at his mother's home in Woolsthorpe, Lincolnshire, because the plague epidemic that was then sweeping England had led the University of Cambridge to close its doors and send the students home. It was during this time that Newton saw ripe apples falling off trees in his mother's garden. It probably wasn't as dramatic as stories have it, but it did remind him of a problem that still needed explaining: why things fall down to earth.

He was busy with lots of scientific matters during this period. Take mathematics, for instance. Galileo, Descartes and many other natural philosophers (that is, scientists) had made great strides in developing mathematics as a subject and, even more importantly, in using it to understand the results of their observations and experiments. Newton was an even better mathematician, and he was brilliant in using it in his science. To describe things like movement and gravity mathematically, algebra and geometry are not enough. You must be able to consider very small units of time and movement: an infinitesimal amount, in fact. When examining a bullet fired from a gun, or an apple falling from a tree, or a planet going around the sun, you must focus on the distance it goes in the smallest conceivable moment of time. Many natural philosophers before Newton had seen the problem and thought up various solutions. But Newton, still in his twenties, developed his own mathematical tools to do the job. He called it his method of 'fluxions', from the word 'flux' which means something changing. Newton's fluxions did the kind of computations that we still do in the branch of mathematics now called calculus. By October 1666, when he had finished a paper written just for his own satisfaction, he was the foremost mathematician in Europe, but nobody but Newton knew it. He didn't publish his mathematical discoveries straightaway; instead, he used them, and only eventually shared his methods and results with his acquaintances.

Besides mathematics, Newton began to investigate light. Since ancient times, it had been assumed that sunlight is white, pure and homogeneous (meaning composed of all the same thing). Colours were thought to be caused by modifications of this essentially pure ray. Newton studied Descartes' work on light and repeated some of his experiments. He used lenses and then a glass object, called a prism, which could break up light. He famously allowed a tiny beam of light into his darkened room, through a prism and then on to the wall twenty-two feet (nearly seven metres) away. If light was homogenous, as Descartes and many others had thought, the projection on the wall ought to be a white circle, the same shape as the hole through which it had passed. Instead, the light appeared as a wide multicoloured band. Newton hadn't exactly made a rainbow, but he was on the way to explaining how they are formed.

During these plague years, Newton also pushed forward with his work on mechanics: the laws governing bodies in motion. We have seen how Galileo, Kepler, Descartes and others had developed ideas to explain (and write out mathematically) what happens when a cannon-ball is fired, or the earth moves around the sun. Robert Hooke, too, had been interested in this. Newton read the writings of these men, but he also went further. He once wrote to Hooke, 'If I have seen further it is by standing on the shoulders of giants.' Do you remember riding on your parent's shoulders? Suddenly being twice or three times as tall reveals all sorts of things you couldn't see by yourself. And that is what Newton was getting at. His wonderful image describes how each scientist, and each generation of scientists, can benefit from the insights of those who came before. This is the essence of science.

But Newton was also himself a giant, and he knew it. The problems arose when Newton didn't feel that others recognised this. Newton's troubles with Robert Hooke began when Newton offered his very first paper to the Royal Society. The Society did what good scientific journals still do today: they sent it to another expert to comment on. We call this 'peer review', and the process is part of the openness that scientists pride themselves on. The Royal Society chose Hooke to read the paper since he, too, had investigated light. Newton did not like Hooke's comments at all, and even wanted to resign as a Fellow of the Royal Society. The Society quietly ignored his letter of resignation.

Following his amazing burst of creative energy in the 1660s, Newton turned his attention to other matters, including alchemy and theology. As always, he kept careful notes on his reading and experiments, which are still being read by people who want to understand this side of Newton's thinking. At the time, he kept these thoughts and investigations fairly quiet, especially his religious views, which differed from the doctrines of the Church of England. Cambridge University required its students to agree to the Church's beliefs. Fortunately for Newton and for science, he had powerful supporters at the university, so he was able to become a Fellow of Trinity College, and later was even elected Lucasian Professor of Mathematics, without ever having to swear that he believed in all the Church's doctrines. He held this professorship for more than twenty years. Unfortunately, he was a terrible teacher, and his students couldn't understand what he was talking about. Sometimes, when he arrived, there was nobody to lecture to. He always talked about respectable subjects like light and motion, not the alchemy and theology that he was secretly pursuing – perhaps those would have been more exciting for his students!

By the mid-1680s, Newton's research into mathematics, physics and astronomy was becoming known. He had written many papers and published a few, but he often remarked that his scientific work was just for himself alone, or for those who came along after his death. In 1684, the astronomer Edmund Halley visited Newton in Cambridge. (Look out for Halley's Comet, named after Edmund Halley, in 2061 when it is next due to be visible from earth.) Halley and Hooke had been discussing the shape of the path taken by one object orbiting around another (such as the earth around the sun, or the moon around the earth). They wondered if gravity would affect the object's path, acting by what we now call the 'inverse square law'. Gravity is only one of several examples of this law. It means that the force of gravity decreases by the square of the distance between the two objects, and of course, increases in the same ratio as they get closer together. The attraction will be mutual, but the mass of the two objects is also important. If one object – say, the earth – is very large, and the other – say, an apple – is very small, the earth will do almost all the attracting. Chapter 12 explained how Galileo used a 'square' function in his work on falling bodies. We will also see it in later chapters, for Nature does seem to like things to happen as a function of something squared, whether it be time, acceleration or attraction. When you're working with squares ($3 \times 3 = 9$, or 3^2, for example), remember that Nature might be smiling.

Halley's visit made Newton put aside his theology and alchemy. He set to work and produced his greatest book, one of the most important in the history of science, even if it is not an easy read. Today it is known as the *Principia* but its full Latin title (Newton wrote it in Latin) is *Philosophiae naturalis principia mathematica* ('Mathematical principles of natural philosophy': remember 'natural philosophy' was the old name for science). Newton's book gave the full details of how his new mathematics could be applied, and explained many aspects of physical nature in numbers rather than wordy descriptions. Only a few people could understand it easily in Newton's lifetime, but its message was appreciated much more widely. It was a new way to see and describe the universe.

Many aspects of Newton's view of the world and the heavens were contained in his three famous laws of motion, which he wrote in the *Principia*. His first law stated that every body either stays at rest or moves in a straight line unless something else – some force – acts upon it. A rock on a mountainside will stay there forever unless something – wind, rain, a human being – causes it to move; and, without any disturbance ('friction'), it would move in a straight line forever.

His second law stated that if something is already moving, a force can change its direction. How great the change depends on the strength of the force, and the change of direction occurs along a straight line, in the direction of the new force. So, if you swat a falling balloon from the side, it will move sideways; if you swat it from above, it will go down more quickly.

His third law of motion concluded that for any action, there is always an equal and opposite reaction. This means that two bodies always act upon each other equally but in opposite directions. You can swat a balloon, and it will move away from your hand, but it will also deliver an action on your hand (you will feel it). If you swat a large boulder, the boulder won't move, but your hand may bounce back, and it will be sore. This is because it is harder for lightweight objects to influence heavy ones than vice versa. (We saw that it was the same with gravity.)

These three laws brought together the puzzles of earlier natural philosophers. In Newton's hands, they explained many observations, from the movements of the planets to the trajectory of an arrow shot from a bow. The laws of motion made it possible to view the whole universe as a giant, regular machine, like a watch that keeps time because of its springs, levers and movements. Newton's *Principia* was recognised as a work of great power and genius. It turned this reclusive, troubled man into something of a celebrity. His reward was a well-paid post as the Warden of the Mint, the place where the government made its coins and regulated the country's supply of money. Newton threw himself into this new job with great gusto, tracking down counterfeiters and overseeing the nation's money supply. He had to move to London, so he resigned all his Cambridge connections and spent the last thirty years of his life in the capital, becoming President of the Royal Society.

During his London years, Newton significantly revised the *Principia*, including some of his further work, as well as answering various criticisms that people had raised since its publication. Scientists often do this. Not long after Robert Hooke died, Newton published his second major scientific work, *Opticks* (1704), about light. Newton and Hooke had quarrelled a lot over which of them had done what first and how to understand the results of their experiments into what light was and how it behaved. Newton had done much of the work for this book nearly forty years earlier, but he had been reluctant to publish it while Hooke was alive. Like *Principia*, *Opticks* was very important. We'll encounter some of its conclusions in later chapters, when other scientists were standing on Newton's shoulders.

Newton was the first scientist to be knighted, becoming Sir Isaac. He enjoyed power but not much happiness. He was not what one would call a nice man, but he was a great one, one of the most truly creative scientists who has ever lived, celebrated for the amazing contributions he made to our understanding of the universe. Newton's *Principia* was the highpoint of the astronomy and physics that had been so actively pursued by Kepler, Galileo, Descartes and many others. In his book, Newton brought the heavens and earth together in one system, for his laws applied throughout the universe. He offered mathematical and physical explanations of the way the planets move and the way bodies fall towards the earth. He provided the foundations of physics that scientists used until the twentieth century, when Einstein and others showed that there was more to the universe than even Sir Isaac had imagined.

CHAPTER 17

Bright Sparks

Have you ever wondered exactly what a flash of lightning is, and why a rumble of thunder follows? Violent displays of thunder and lightning happen high up in the sky, and are pretty dramatic, even if you know what causes them. Just as bolts of lightning always seek earth, by the early eighteenth century scientists had started to puzzle over this and about electricity much closer to home.

Another puzzle was over what came to be known as magnetism. The ancient Greeks knew that if you rub amber (a yellowish semi-precious stone) very hard, it attracts small nearby objects to it. The cause of this power was difficult to understand. It seemed different from the constant power of a different kind of stone – the lode-stone – to attract objects containing iron. Just as a lodestar is a star that shows the way (especially the Northern Star), so the lodestone also guided travellers: it was a piece of a special mineral that, if suspended so that it could swing freely, would always point towards the magnetic poles. Lodestones could also be used to magnetise needles, and by the time of Copernicus, in the mid-sixteenth century, crude compasses were being used by seamen to help find their direction, since one end of the moveable needle of the compass always pointed to the north. An English doctor named William Gilbert wrote about this in 1600, when the word 'magnetism' had arrived. Both electricity and magnetism could produce entertaining effects and were popular topics for scientific lectures as well as after-dinner games.

Soon, people obtained even more powerful effects by rotating a glass globe on a point and rubbing it as it turned. You could feel and even hear the sparks as they were produced on the glass. This device became the basis of what was called the Leyden Jar, named after the town in the Netherlands where it was invented, in about 1745, by a professor at the university. The jar was half filled with water and connected to an electricity-generating machine by a wire. The connecting piece was called the 'conductor' because it allowed the mysterious power to pass into the water in the jar, where it was stored. ('To conduct' means 'to lead'.) When a laboratory assistant touched the side of the jar and the conducting piece, he got such a jolt that he thought it was all over for him. The report of this experiment caused a sensation and Leyden Jars became all the rage. Ten monks once linked hands and when the first one touched the jar and conducting piece, they were all jolted simultaneously. An electric shock, it seemed, could be passed from person to person.

What exactly was going on? Beyond the games, there were serious scientific issues at stake. There were a lot of theories flying about, but one man who brought some order to the subject was Benjamin Franklin (1706–90). You might know him as an early American patriot who helped write the Declaration of Independence (1776) after the United States successfully achieved its independence from the British Empire. He was a witty, popular man, full of homespun wisdom, such as 'Time is money', and 'In this world, nothing can be said to be certain, except death and taxes'. Next time you sit in a rocking chair, or see someone wearing bifocal glasses, think of him: he invented them both.

Largely self-taught, Franklin knew a lot about a lot of things, including science. He felt equally at home in France, Britain and America, and he was in France when he performed his most famous scientific experiment, with lightning. Like many people in the 1740s and 1750s, Franklin was curious about Leyden Jars and what they could show. In his hands, they showed far more than had been thought. First, he realised that things could carry either positive or negative charges – as you can see marked by the '+' and '–' at the opposite ends of a battery. In the Leyden Jar, the connecting wire and the water inside the jar were 'electrised positively, or plus', he said, while the outer surface was negative. The positive and negative were the same strength and so cancelled each other out. Further experiments convinced him that the Jar's actual power lay in the glass, and he made a kind of battery (he invented the word) by placing a piece of glass between two strips of lead. When he connected his device to a source of electricity, this 'battery' could be discharged of its electricity. Unfortunately, he did not pursue this discovery any further.

Franklin was not the first to puzzle about the relationship between the sparks generated by machines on earth and sparks in the sky, that is, lightning. But he was the first to apply what he had learned about the Leyden Jar to try to see how they might be connected. He devised a clever (but dangerous) experiment. He argued that electricity in the atmosphere would collect on the edge of clouds, just as it did in the Leyden Jar. If two clouds collided with each other, as they rolled across the sky during a thunderstorm, there would be a discharge of electricity – a flash of lightning. By flying a kite during such a storm, he could show that his idea was correct. The person flying the kite needed to be properly insulated from the electricity (by using a wax handle to hold the kite-string) and 'grounded' (with a piece of wire attached to him and trailing on the ground). Without these precautions, the shock of the electricity might kill him, and indeed, one unfortunate experimenter did die because he didn't follow Franklin's instructions. The kite experiment convinced Franklin that the electricity of lightning was like the electricity of Leyden Jars.

First gravity, now electricity: things in the heavens and on earth were being brought ever closer together.

Franklin's work on electricity had immediate practical consequences. He showed that a metal pole with a sharp point conducted electricity to the ground. So, if such a pole were placed on the top of a building, with an insulated conducting body leading from it all the way down to the earth, lightning would be conducted away from the building, and it would not catch fire if struck by the lightning. This was a serious problem when houses were built mostly of wood and sometimes had thatched roofs. Lightning rods, as they are still called, act on this principle, and even now we use the word 'earth' for the bit of insulated wire in our electric plugs that takes away excessive electrical charge in things like washing machines and refrigerators. Franklin connected a lightning rod on his own house, and the idea caught on. There were important results to be had from understanding electricity.

The study of electricity was one of the most exciting areas of scientific research in the eighteenth century, and many 'electricians', as they were called, contributed to what we know today. Three in particular have left their names with us. The first was Luigi Galvani (1737–98), a doctor who liked to tinker with electrical apparatus and animals. He practised medicine and taught both anatomy and obstetrics (the medical management of childbirth) at the University of Bologna, but he was also much interested in physiological studies. While investigating the relation between muscles and nerves, he discovered that a frog's muscle could be made to contract if the nerve attached to it were connected to a source of electricity. After further research, he likened muscle to a Leyden Jar, able to generate and discharge a current of electricity. Electricity was an important part of animals, Galvani said. Indeed, 'animal electricity', as he termed it, seemed to him to be an essential ingredient of how animals functioned. And he was right.

Static electric shocks, which occur when electricity that has built up on the surface of an object is discharged, are still called galvanic shocks. Scientists and electricians use galvanometers to measure electric currents. Galvani's notion of animal electricity attracted a good deal of criticism, especially from Alessandro Volta (1757–1827), a scientist from Como in northern Italy. Volta had a low opinion of doctors who branched out into physics, and he set out to show that animal electricity did not exist. Volta and Galvani had a very public debate about the interpretation of Galvani's experiments. In the course of his extensive work aimed at discrediting Galvani, Volta examined the electric eel, which, as could be demonstrated, did produce electricity. He believed that even these animals did not make Galvani's 'animal electricity' more convincing. More importantly, Volta discovered that if he built up successive layers of zinc and sliver, and separated them by layers of wet cardboard, he could produce a continuous electric current through all the layers. Volta sent news of his invention, which he called a 'pile', to the Royal Society in London. Like the Leyden Jar, it created a sensation in England and France.

At this time, France was busy conquering northern Italy, and the French Emperor, Napoleon Bonaparte, decorated the Italian physicist for his invention, for it offered a reliable source of electric currents for experimental research. Volta's 'pile' went on to play an essential part in early nineteenth-century chemistry. It was the practical development of Franklin's 'battery', and has become essential in our lives today. We remember Volta because his name gave us the word 'volt' which is one way we measure electrical power – check out the packaging next time you change a battery.

Our third great electrician (and formidable mathematician) also gave his name to the measurement of electricity: André-Marie Ampère (1775–1836). We get the word 'amp' from his name. Ampère lived through the trauma of the French Revolution and its aftermath, during which his father lost his head on the guillotine. His personal life was equally sad. His beloved first wife died after the birth of their third child, and his second marriage was deeply unhappy and ended in divorce. His children turned out badly, and he was constantly beset with money worries. In the middle of this chaos, Ampère realised some fundamental things about mathematics, chemistry and, above all, what he called 'electrodynamics'. This complicated subject brought together electricity and magnetism. Despite its complexity, Ampère's simple but elegant experiments showed that magnetism was in fact electricity in motion. His work underpinned that of Faraday and Maxwell, and so we will talk about it in more detail when we come to these later giants of electromagnetism. Although later scientists showed that many of the details of Ampère's theories did not lead anywhere, he provided the starting point for much research into electromagnetism. It is important to remember that science is also about sometimes getting things wrong.

By the time of Ampère's death, electricity had gone a long way towards being tamed. Franklin's work had been homespun and, important though it was, he was an ingenious amateur compared with Galvani, Volta and Ampère, who used more sophisticated equipment, and worked in laboratories. Galvani had the last laugh on Volta, for we now know that electricity plays an important part when muscles and nerves interact.

CHAPTER 18

The Clockwork Universe

The American Revolution (also known as the American War of Independence) in 1776, the French Revolution in 1789, and the Russian Revolution in 1917 each swiftly brought about new forms of government and a new social order. There was also a Newtonian Revolution. Fewer people have heard of the Newtonian Revolution, but it was just as important, and although it took decades rather than years to work its effect, its consequences were profound. The Newtonian Revolution described the world in which we live.

After he died in 1727, Sir Isaac continued to be a towering figure in the eighteenth century. In every field of endeavour, people wanted to be the 'Newton' of their subject. Adam Smith wanted to be the Newton of economics; some called William Cullen the Newton of medicine; Jeremy Bentham strove to be the Newton of social and political reform. What they all sought was a general law or principle that would glue their discipline together, just as Newton's gravity seemed to hold the universe in its regular and stately progression through the seasons and years. As the poet Alexander Pope joked, 'Nature, and Nature's laws lay hid in night./ God said, *Let Newton be!* and all was light.'

As an Englishman, Pope might have been biased in favour of his countryman. In France, Germany and Italy, Newton cut a sizeable figure even in his own lifetime, but there were other scientific traditions that still counted. In France, Descartes' mechanical vision of the universe remained powerful. In Germany there were squabbles over who had invented calculus, with the admirers of the philosopher G.W. Leibniz (1646–1716) insisting that Newton was less important in developing this mathematical tool than their man. In Britain, however, Newton attracted many followers, who were only too pleased to call themselves 'Newtonians', and who used his magnificent insights in mathematics, physics, astronomy and optics.

Gradually, however, the power of Newton's experimental optics and laws of motion also took hold of European thought. His reputation was helped by a most unlikely advocate: the poet, novelist and man of letters Voltaire (1694–1778). Voltaire's most famous creation was the loveable character, Candide, who featured in an adventure story. Candide lives a life of continuous disaster – everything that can go wrong does go wrong – but he never forgets his philosophy: the world that God has created *must* be the best possible one. So he remains cheerful, sure that what happens to him, no matter how dreadful, is for the best 'in this best of possible worlds'. (After his horrible adventures he decides that he should have stayed at home and tended his garden: pretty good advice, actually.)

Candide was a gentle dig at the philosophy of Newton's rival in the invention of calculus, Leibniz. Voltaire was a great fan of Newton and, in fact, all things English. He spent a couple of years in England and was very impressed with the freedom of speech and thought there. (Voltaire was imprisoned at home in France for criticising the Catholic Church and the French king, so he knew how important free speech is.) He also came away from England appreciating Newton's achievements, and he wrote a popular version of

Newton's ideas for ordinary people in French. Voltaire's book found many readers in Europe, where everyone was discussing the ways in which Newton's mathematics and physics made sense of the movements of the planets and stars, the daily ebb and flow of the tides, the trajectory of bullets, and of course the falling of apples.

Newton gradually acquired his towering reputation because the tools – both mathematical and physical – that he set out in his famous *Principia* actually worked. These tools helped mathematicians, physicists and astronomers to study a number of problems that Newton had only touched on. No work of science is ever the last word, and so it was with Newton's. Many individuals were happy for Newton to be the giant on whose shoulders they could stand. And in many instances, he did help them see further.

Let's look at three examples: the causes of the tides, the shape of the earth, and the number and orbits of the planets in the solar system.

There are low and high tides: a low tide is when the sea is 'out' and you have to walk a lot further before you can have a swim, and a high tide is when the sea is 'in' and it's washing away your sandcastles. The tides have a regular, daily pattern, and knowing about them was important for sailors, who might need a high tide to get the ship into harbour. Aristotle had drawn a connection between the tides and the moon. After it became common to believe that the earth actually moves, some compared the tides with the waves you can make in a bucket of water by tilting it to and fro. For Newton, gravity was the key. He argued that the moon's 'gravitational pull' is greatest when the moon is closest to the earth. (Like the earth revolving around the sun, the moon revolves around the earth in an ellipse, so the distances between the earth and moon vary regularly.) The gravity of the moon attracts the water in the oceans towards it. As the earth revolves, an area of sea will become nearer to, and then further away from, the moon, and so the increasing and decreasing force of gravity helps raise and lower the oceans in the regular fashion that we can see. This explains the high and low tides. Newton was right to think that the tides illustrated gravity in action.

Later Newtonians refined the master's calculations. The Swiss doctor Daniel Bernoulli (1700–82) offered a closer analysis of tides in 1740. He was much more interested in mathematics, physics and navigation than in medicine, and he also helped explain how strings vibrate (as when you strum a guitar) and how pendulums swing (as in grandfather clocks). He improved the design of ships, too. At the medical school in Basel, he used Newtonian mechanics to look at things such as the way our muscles contract and shorten to make our limbs move. His work on tides was in response to a question set by the Academy of Sciences in Paris, which offered a prize for the best answer – learned societies often did this. Bernoulli shared the prize with several others, each helping to explain why tides behave as they do, and including, in their explanations, the effect of the gravitational pull of the sun as well. When two things, like the earth and the moon, attract each other, the mathematics is relatively simple. In the real world, the sun, planets and other things having mass complicate the picture, and the mathematics becomes much more difficult.

The Paris Academy of Sciences was also involved with a second major question of Newtonianism: was the earth a round ball? It was easy to see that it wasn't completely smooth, like a table tennis ball – there were mountains and valleys. But was it basically round? Newton had said no, since he had shown that the force of gravity at the equator was slightly different from the force of gravity in northern Europe. He knew this by experiments with a pendulum. The swing of a pendulum is influenced by the force of the earth's gravity; the stronger the gravity, the faster the pendulum moves, and so it takes a shorter amount of time for it to complete its to-and-fro cycle. Sailors had measured how far the pendulum swung in exactly one second, and the distance was slightly shorter at the equator. This difference told Newton that the distance to the centre of the earth was slightly greater at the equator. If the earth were a perfect ball, it would be the same distance everywhere from the surface to the centre. Consequently, Newton said that the earth was actually flattened at the poles – as if it had been squashed from top to bottom – and it bulged a bit at the equator. He thought this shape had been created by the earth's rotation on its north–south axis when it was still very new and cooling from its fluid state. Newton hinted that this meant that the earth was older than 6,000 years, but he never revealed how old he thought the earth really was.

When Newton's work was being debated in France during the 1730s, many French scientists refused to believe that the earth had this imperfect shape. So Louis XV, King of France, sent out two expeditions, one to Lapland, near the Arctic Circle, and one to Peru, near the equator – an expensive way to test a simple fact. What the two expeditions did was to measure the precise length of one degree of latitude at these two locations. Latitude is a measure of the north–south axis of the earth, with the equator being zero degrees, the North Pole +90 degrees and the South Pole –90 degrees. (It takes 360 degrees to go around the globe completely.) You can see the lines of latitude drawn from side to side across a map of the world. If the earth were perfectly round, each degree of latitude would be the same. The Lapland expedition returned first (they hadn't had to travel so far) but when the Peru group came back, after nine years, it was shown that the degree of latitude in Lapland was longer than the one in Peru, exactly as predicted by Newton's model. These results helped raise Newton's reputation in continental Europe.

Astronomers all over Europe were looking at the stars and planets in an attempt to predict how they moved, and therefore where they would be observed each evening (or each year). These predictions became ever more precise, as more and more observations were done, and as the mathematical analysis of their movements became more accurate. Building bigger telescopes enabled astronomers to see further into space, and then to discover new stars, and even new galaxies. One of the most important of these stargazers was a refugee to England from Germany, William Herschel (1738–1822). Herschel was a musician, but his passion was looking at the heavens. One night, in 1781, he noticed a new object, which was not a star. At first he thought it was probably a comet and he described it to a local group in Bath, where he lived. His observation attracted the attention of others, and it quickly became clear that Herschel had discovered a new planet. It was eventually named Uranus, after a character in Greek mythology.

This discovery changed Herschel's life and enabled him to devote himself entirely to astronomy. King George III, whose family had also come from Germany, took an interest in Herschel's work. George helped Herschel build the world's largest telescope, and eventually to come to live near Windsor, where one of the royal castles was located. So devoted was Herschel to looking at the heavens, that when he moved to Windsor he arranged his life so that he need not miss a single night's observations. In all his work, Herschel was helped by his sister Caroline (1750–1848), who was also an expert astronomer. Herschel's son John (1792–1871) also continued his father's work, making it a family business.

William Herschel not only looked at stars, planets and other heavenly bodies, but he also thought deeply about what he was seeing. Since he had the best telescopes of his time, he could see further. He produced catalogues of stars that were much larger and more accurate than ever before. He realised that our galaxy, the Milky Way, was not the only galaxy in the universe, and he puzzled long and hard about what were called 'nebulae', areas in the sky that appeared as fuzzy white blotches. A few of these can sometimes be seen on a clear night with the naked eye, but Herschel's telescope revealed many more of these blotchy areas. The Milky Way begins to look fuzzy as we peer at its more distant points, and astronomers had assumed that nebulae were simply clusters of stars. Herschel showed that some of them probably are, but that others were enormous areas of gaseous clouds swirling around in deep space. In addition, by looking at 'double stars', pairs of stars close to each other (well, it's 'close' considering the distances we are talking about), he showed that the behaviour of these stars could be explained by gravitational attraction: Newton's gravity was shown to extend even into the outer reaches of space.

Newton's laws of gravity and of motion, along with his mathematical analysis of force (power), acceleration (increasing speed), and inertia (the tendency to keep moving in a straight line), became the guiding principles for natural philosophers during the eighteenth century. No one did more to show how much these principles could explain than the Frenchman, Pierre Simon de Laplace (1749–1827). Laplace worked with Lavoisier, whom we'll meet in Chapter 20, but, unlike his unlucky friend, Laplace came through the French Revolution unharmed. Admired by Napoleon, he was a leading figure in French science for half a century. Laplace used Newton's laws of motion and his mathematical tools to show that the things one could see in the sky could be understood, and that future movements of planets, stars, comets and asteroids could be predicted with accuracy. He developed a theory of how our solar system, with the sun and its planets, could have been born millions of years ago from a vast explosion, with the sun throwing off great chunks of hot gases that gradually cooled to form the planets (and their moons). He called this the 'nebular hypothesis', and he offered some very complicated mathematical calculations to show that it might have happened that way. Laplace was describing a version of what we now call the Big Bang (Chapter 39), although physicists today know a great deal more about this than Laplace could have known.

Laplace was so impressed with the power of Newton's laws of motion that he believed that if we could only know where every particle in the universe was at a given moment, we could predict the running of the whole universe to the ends of time. He realised that it was not possible to do this. What he meant was that the laws of matter and motion were such that the whole universe really did work like a very well-made clock, and that it kept perfect time. His vision of a clockwork universe served scientists for a century after him.

CHAPTER 19

Ordering the World

Our planet is home to a bewildering variety of plants and animals. We still don't know exactly how many insects or sea creatures there are. We rightly worry that the human race is reducing their number. 'Endangered species', such as giant pandas and Indian tigers, are in the news almost every day. For us as concerned human beings, the important word in 'endangered species' is *endangered*, but for scientists, an equally significant word is *species*. How do we know that the giant panda is not the same kind of animal as the grizzly bear, or the wildcat different from the pet cat we stroke?

Adam, in the Bible's Book of Genesis, is given the job of naming the plants and animals in the Garden of Eden. All human groups have some way of organising the living world around them. All languages have names for the plants and animals people use, whether they are cultivated, gathered or provide transport, meat, hides or milk.

During the seventeenth and eighteenth centuries, European explorers began to bring back many new kinds of plants and animals from exotic parts of the world: from North and South America, Africa, Asia, and then Australia and New Zealand, as well as islands in the world's oceans. Many of these new creatures were wonderfully different from the familiar plants and animals of the Old World, but when they were closely examined, a lot of them were not all *that* different. For example, elephants found in India and Africa were so similar that the same name seemed appropriate. Yet, there were small differences. How should we account for these slight differences, and for the rich variety of nature?

From Antiquity, there had been two basic answers to this question. One was to assume that nature was so bountiful that it was not surprising that many, many new kinds of plants and animals were being found in remote parts of the world. These new discoveries were thought to be simply filling the gaps in what naturalists called the 'Great Chain of Being', an idea we met way back in Chapter 5. Those who believed in the Chain of Being argued that God was so powerful that he created every creature that could possibly exist. They weren't surprised to find animals that combined characteristics of other animals, like whales and dolphins in the oceans, which looked like fish, but breathed and gave birth like land animals; or bats, which looked like birds in that they had wings and flew, but didn't lay eggs. This was because these naturalists thought all the curious aspects of plant and animal life could be explained as being part of the Chain of Being. The idea of the 'missing link' in this chain, which you might have heard about when an important new fossil is found, has been around for a long time.

The second answer was to assume that God originally created each *kind* of plant and animal, and that the vast variety of nature we can see around us is the result of generation after generation producing their young. Oak trees produce saplings from their acorns, just as cats give birth to kittens, which grow up to have more kittens, and so on. And with each generation, or hundreds of generations, or thousands, the trees and cats would become more diverse. That is, the vast variety of nature was to be understood as being caused by changes that happened over time, though each plant or animal could still be said to relate to an original design. To map out all of the original plants and animals would display God's plan, as a 'tree of life'.

During the eighteenth century, two naturalists dominated thinking on these issues, and they happened to reflect these two differing approaches. The first was a French nobleman, the Comte de Buffon (1707–88). Georges-Louis Leclerc, a rich man, devoted his life to science. He spent part of the year on his estate and the other part in Paris, where he was in charge of the king's gardens – these were much like a zoo or wildlife park today. Early on, he was a great admirer of Newton and his physics and mathematics, but most of his long life was spent investigating the natural world. His aim was to describe the earth and all the plants and animals on it. All his careful research was collected in a massive work of 127 volumes, called simply *Histoire naturelle* ('Natural history'). At that time, 'history' also meant 'description', and in these books Buffon set out to describe all the animals (and a few plants) that he could get hold of.

Buffon described nearly everything he could about his animals: their anatomy, the way they moved, what they ate, how they reproduced, what uses they were to us, and much else besides. It was a wonderfully modern attempt to see animals in their environment as far as possible. In one volume after another, he examined many of the known mammals, birds, fishes and reptiles. This massive work came out over about forty years, from 1749, and readers eagerly awaited each new volume. They were translated into most European languages.

Buffon was fascinated by all the characteristics of each animal he examined. As he famously said, 'Nature knows only the individual', meaning that there was no order in nature, only a lot of individual plants and animals. It was only humans who tried to classify them into groups, for their own use. Of the Great Chain of Being, he said that nature was very full but it could only be studied one creature at a time.

Buffon's great rival was the Swedish doctor and naturalist, Carl Linnaeus (1707–78). Linnaeus learned medicine but his real passion was plants. He spent most of his life as a professor at the University of Uppsala, in northern Sweden. Here he maintained a botanical garden, and sent many students all over the world to collect plants and animals for him. Some of his students died on their travels, but his followers remained devoted to Linnaeus's great goal: to name accurately all the things that exist on earth. To help with his naming, Linnaeus *classified* them, that is, he defined their essential characteristics. This allowed him to place them within the 'order of nature'. When he was still in his twenties, in 1735, he produced a short book called *Systema Naturae* ('The system of nature'). The book was basically a long list of all the known species of plants and animals, grouped by genera. He published twelve editions in his lifetime, each time expanding his list as he learned about more kinds of plants and animals, especially those that his students discovered for him in America, Asia, Africa and other parts of the world.

Since the ancient Greeks, naturalists had asked whether there could be a 'natural' classification of the things in the world. Do things have a timeless or God-given relationship to each other? And if so, how can we find it out? In the Christian era, the most common assumption was that God had created each species of plant and animal 'in the beginning', for Adam to name, and that what we see now was the product of time and chance.

Linnaeus was sympathetic to this view, but he realised just how much plants and animals had changed since their creation. This made a 'natural' classification very difficult to achieve. So what was needed first, he thought, were some simple rules to order and classify all the things of the world. Second, he wanted to give things a simple label to identify them. This was his life's task: he saw himself literally as a second Adam, giving things their precise names. After all, how could zoologists or botanists discuss a kind of 'dog' or a kind of 'lily' unless they knew exactly what kind they were talking about? Nature, Linnaeus thought, had to have pigeonholes, and when everything was in its proper box, then science could be done.

Linnaeus classified just about everything: minerals, diseases, plants and animals. Among the animals, he made a bold move: he included human beings in his scheme. In fact, he gave us the biological name we still have: *Homo sapiens*, which literally means 'wise, or knowing, man'. Many naturalists before Linnaeus had confined themselves to what is sometimes called the 'natural world', and therefore excluded human beings from their schemes. Linnaeus, the son of a minister, was deeply religious. As he pointed out, however, there were no biological reasons why human beings were not simply animals, as are dogs and monkeys, and so they needed to be included in his system of nature.

The two most important categories for Linnaeus in his work of *taxonomy* (the scientific word for classification) were the *genus* and the *species*. He always used a capital letter to name the genus (we still do), and a lower-case letter for the species: thus *Homo sapiens*. The genus was a group of plants or animals that shared more basic characteristics than species share. For example, there are several different species of cat in the genus *Felis*, including our domestic cat (*Felis catus*) and the wildcat (*Felis silvestris*). (In those days everyone learned Latin in school, so his labelling would have been easy to understand: *felis* meant 'cat', *catus* 'cunning', and *silvestris* 'of the woods'.)

Linnaeus knew that there were different levels of resemblance or difference between living creatures. At the top of his grand scheme he had three *kingdoms*: plants, animals and minerals. Under these were *classes*, such as the vertebrates (animals with spinal cords: donkeys, lizards, and so on); within a class were *orders*, such as the mammals (creatures that suckle their young); one notch down was the *genus*; followed by the *species*. Below species, there were *varieties*. Within the human species, these varieties were called 'races'. Of course, there are individuals – a person, plant or animal with its own peculiar characteristics, such as height, male or female, hair or eye colour, or tone of voice. But you don't classify individuals as such, rather you put them into a group that you can then classify. Later scientists found they had to add extra ranks into Linnaeus's original system, such as families, sub-families and tribes. Lions, tigers and domestic cats are now all grouped in the *family* of cats.

The sum total of all individual plants and animals makes up the living world, and it was this that Buffon referred to when he insisted that this basic category – the individual – was the only certain one.

The really crucial level for Linnaeus was that of the species. He devised a simple system for identifying each plant species, based on the male and female parts of their flowers. It allowed amateur botanists to roam the woods and fields and identify what they were seeing. Even though it was only in plants, Linnaeus's sexual system disturbed some people and also stimulated a few mildly erotic poems. Most importantly, his classification of plants worked well. It gave botany a real boost. After Linnaeus's death, his important plant collections were bought by a wealthy Englishman, who established the Linnean Society of London. It is active to this day, after more than 200 years.

We still use many of the names that Linnaeus introduced to identify plants and animals. One of them was the order of animals that includes human beings, the *primates*. We share that order with apes, monkeys, lemurs and other animals that share many characteristics with us. Linnaeus did not believe that one species can evolve into another: he believed that God had specially created each separate species of plant and animal. But he realised that human beings were part of nature, and that the rules by which we study the natural world could also be used to understand mankind. What we mean exactly when we say that this or that group of plants or animals is a biological species continued to puzzle naturalists. It still does. But Linnaeus's framework was changed a century later, by another naturalist who also loved plants: Charles Darwin. We will pick up the story in Chapter 25.

CHAPTER 20

Airs and Gases

'Air' is a very old word. The word 'gas' is much newer, only a few hundred years old, and the shift from air to gases was crucial. For the ancient Greeks, air was one of the four fundamental elements, just one 'thing'. But Robert Boyle's experiments in the seventeenth century had challenged this view, and scientists had come to realise that the air that surrounds us, and that we all breathe, is made up of more than one substance. From then on it was much easier to understand what was happening in many chemical experiments. Lots of experiments produced something that bubbled up, or went up in a puff and then disappeared into the air. Sometimes the experiment seemed to change the air: chemists often produced ammonia, which made their eyes water, or hydrogen sulphide, which stank of rotten eggs. But without being able to collect the gases in some way, it was hard to know what was going on. Isaac Newton had showed that measurement was important, but it was hard to measure a gas if it was just loose in the atmosphere.

So chemists had to find ways to collect pure gases. The most common way of doing this was to conduct the chemical experiment in a small closed space, like a sealed box. This enclosed space was then connected by a tube to an upside-down container completely filled with water. If the gas didn't dissolve in the water – and some gases do – it could bubble up to the top and push the water down. Stephen Hales (1677–1761), an ingenious clergyman, devised a very effective 'water bath' for collecting gases. Hales spent most of his long life as the vicar of Teddington, then a country village, now swallowed up into London. A modest and retiring man, he was also extremely curious and a constant experimenter. Some of his experiments were pretty horrible: he measured the blood pressure in horses, sheep and dogs by directly sticking a hollow tube into an artery. This was attached to a long glass tube, and he simply measured how high the blood rose, which equalled the blood pressure. For a horse, the glass tube had to be nine feet tall (2.7 metres) to prevent the blood spurting out the top.

Hales also studied the movement of sap in plants and measured the growth of the different parts of plants. He painted tiny specks of ink at regular intervals on their stems and leaves, and then recorded the distances between the specks before and after the plant had grown. He showed that not all the parts grew at the same rate. Hales then used his apparatus for collecting gases to see how plants react in different conditions. He saw that they were using 'air', as the atmosphere was still called. (In 1727 his book *Vegetable Staticks* laid the foundations for the later discovery of photosynthesis, which is how plants use sunlight as a source of energy, and are able to change carbon dioxide and water into sugars and starches, and 'breathe' out oxygen. It is one of the most fundamental processes on our planet. But we are getting ahead of ourselves, and at that stage no one knew about oxygen.)

Remember the word *pneuma*, from Chapter 6? 'Pneumatic' just means 'relating to air', and pneumatic chemistry – the chemistry of airs – was one of the most important areas of science in the eighteenth century. (Did you notice that 'airs' was plural there?) Pneumatic chemistry was where it was at from the 1730s onward. It was not just that the older notion of 'air' was giving way to the much more dynamic idea of it actually being made up of several kinds of gases. Scientists were also discovering that most substances can exist as – or be transformed into – a gas, given the right conditions.

Stephen Hales had led the way with his water bath, and his demonstration that plants, as well as animals, need air. This 'air' was understood to be a gas that was released when something burned. A Scottish doctor and chemist, Joseph Black (1728–99), collected this 'air' (which he called 'fixed air') and showed that while plants could live in it and use it, animals would die if they were placed in a container with just fixed air to breathe. They needed something else. Black's 'fixed air' is now called carbon dioxide (CO_2), and we know it's an essential part of the life cycles of plants and animals. (It is also a 'greenhouse gas', a main cause of the 'greenhouse effect', which is leading to global warming.)

A reclusive aristocrat, Henry Cavendish (1731–1810), spent his days in his private laboratory in his London house, experimenting and measuring. He discovered more about fixed air, and collected another air, one that was very light, and exploded when sparked in the presence of ordinary air. He called it 'inflammable air'. We now call it hydrogen, and it turned out that the explosion produced a clear liquid that was nothing other than water! Cavendish also worked with other gases, such as nitrogen.

No one was as successful in pneumatic chemistry research as Joseph Priestley (1733–1804). Priestley was remarkable. A clergyman, he wrote books on religion, education, politics and the history of electricity. He became a Unitarian, a member of a Protestant group that believed that Jesus was only a very great teacher, not the Son of God. Priestley was also a materialist, teaching that all the things of nature could be explained by the reactions of matter: there was no need for a 'spirit' or 'soul'. During the early days of the French Revolution, which he supported, his house in Birmingham was burned down by people who feared that liberal religious and social views like his might bring revolution across the Channel. He fled to the United States, where he lived the last ten years of his life.

Priestley was also a very busy chemist. He used fixed air to make soda water, so remember him the next time you have a fizzy drink. Priestley identified several new gases, and, like all pneumatic chemists, he wondered what happens when things burn. He knew air played a part in burning, and he also knew that there was a kind of 'air' (a gas) that made things burn even more vigorously than the 'ordinary' air that surrounds us. He made this 'air' by heating a substance that we know as mercury oxide, and collecting the gas in a water bath. He showed that animals could live in it, as plants could in fixed air. Priestley's new 'air' was something special: indeed, it seemed to be the principle that was involved in many chemical reactions, as well as in breathing and burning. He thought it could all be accounted for by a substance called 'phlogiston', and that all things that can burn contain phlogiston, which is released in the burning process. When the air around becomes saturated with phlogiston, they can no longer burn.

Many chemists used this idea of phlogiston to explain what happens when things burn, and why some 'airs' would make things in a closed container burn for a time, and then seem to make them go out. Burn a lump of lead, and the product (what is left behind) will be heavier than the original lump. This suggested that phlogiston, which scientists thought was contained in the lead and was released through burning, must have a negative weight – that is, it makes whatever contains it lighter than things that don't contain it.

When most things burn, the products are gases that are difficult to collect and weigh. Burn a wooden twig, for example, and the product that is esay to see – the ash – is much lighter than the original twig; to get the total weight of the product, the gases given off would have to be collected, weighed and added on.

In Priestley's scheme, phlogiston took the place of what we call oxygen, except that it had almost exactly the opposite properties! For Priestley, when things burned, they lost phlogiston, and became lighter; but we would say they combine with oxygen, and we now know that things get heavier when this happens. When the candle went out in a closed container, or if a mouse or bird died after a while of being sealed inside a closed container with ordinary air, Priestley said it was because the air was saturated with phlogiston; we now know that it's because the oxygen has been used up. It reminds us that it is possible to do very careful experiments, and take careful measurements, but explain the results in very different ways.

The man who named oxygen is still known as the 'father' of modern chemistry. Antoine-Laurent Lavoisier (1743–94) met a violent death during the French Revolution. He was arrested, tried and guillotined, not because he was a chemist, but because he was a 'tax farmer'. In pre-Revolutionary France, rich men could pay a fee to the State to become tax collectors, and then keep what they could collect. The system was rotten, but there is no evidence that Lavoisier abused it. In fact, he spent a lot of his time before the Revolution doing important scientific and technical research for the State, investigating a number of important questions in manufacturing and agriculture. But he was an aristocrat, and the Revolutionary leaders hated him and his class, and he paid the price.

Like Priestley, Cavendish and the other pneumatic chemists, Lavoisier was an enthusiastic experimentalist, and was helped by his wife. In fact Madame Lavoisier was an important figure in science. Marie-Anne Pierrette Paultze (1758–1836) married Lavoisier when she was only fourteen years old (he was twenty-eight), and they worked together in the laboratory, performing experiments, taking readings and recording the results. In addition, Madame Lavoisier was a charming hostess. She and her husband entertained learned men and women who discussed the latest developments in science and technology. Theirs was a happy marriage of real partners.

As a schoolboy, Lavoisier loved science. His sharp mind and scientific ambition were evident from an early age. Like most students who studied chemistry then, he grew up with the phlogiston idea, but he exposed a number of logical and experimental flaws in it. Lavoisier was determined to have the best apparatus available. He and his wife devised new laboratory equipment, always with the aim of improving accuracy in chemical experiments. He used very accurate scales to weigh the substances in his experiments. Several different kinds of experiments convinced him that when things burn, the total weight of all their products increases. This involved collecting and weighing the gases that combustion produced.

Lavoisier also continued to investigate what happens when we (and other animals) breathe. These experiments assured him that the substance involved in both combustion and respiration was a single, real element, and not some kind of substance like phlogiston. This element also seemed to be necessary for acids to form. The chemical reactions of acids and alkalis (the latter are sometimes called 'bases') had long fascinated chemists. Remember Robert Boyle's invention of litmus paper? Lavoisier continued this line of work. Indeed, he believed that oxygen (which means 'acid former') is so important in acids that they always contain that element. We now know that this is not true (hydrochloric acid, one of the most powerful acids, contains hydrogen and chlorine, but no oxygen). Yet much of what Lavoisier said about oxygen is still part of our knowledge today. We now know it is the element needed for things to burn, or for us to breathe, and that those two seemingly different processes have much in common. Humans use oxygen to 'burn', or process, sugars and other things we eat, to give our bodies energy to carry out our daily functions.

Lavoisier and his wife continued with their chemical experiments during the 1780s, and in 1789, just on the eve of the French Revolution, Lavoisier published his most famous book. Its English title is *Elements of Chemistry*, and it is just that. It is the first modern textbook of the subject, full of information on experiments and equipment, and containing his reflections on the nature of the chemical element. We now call an *element* some substance that cannot be broken down any further by chemical experiments. A *compound* is a combination of elements which, given the right experiment, can be broken down. So water is a compound, made up of two elements, hydrogen and oxygen. This distinction was at the heart of Lavoisier's important book. His list of elements, or 'simple substances', did not contain all the elements chemists now recognise, as many had not yet been discovered. It did include surprising things such as light and heat. But Lavoisier laid down the basic framework for understanding the difference between an element and a compound.

Just as important was his belief that the language of chemistry must be precise. With several colleagues, Lavoisier reformed the language of his subject, demonstrating that to do good science, you needed to be precise in the words you use. (Linnaeus would have agreed.) Chemists needed to be able to refer to the compounds and elements they were experimenting with, so that any other chemist, anywhere in the world, would know they were dealing with exactly the same things. He wrote, 'We think only through the medium of words.' After Lavoisier, chemists increasingly shared a common language.

CHAPTER 21

Tiny Pieces of Matter

Atoms used to have a pretty bad name. Remember the ancient Greeks with their notion of atoms as part of a universe that was random and without purpose? So how is it that for us today, being made up of atoms seems so natural?

The modern 'atom' was the brainchild of a thoroughly respectable Quaker, John Dalton (1766–1844). A weaver's son, he went to a good school near where he was born, in the English Lake District. He was especially skilled in mathematics and science, and a famous blind mathematician encouraged his scientific ambitions. Dalton settled in nearby Manchester, a thriving and rapidly growing town during the early Industrial Revolution, when factories began to dominate the making of all kinds of goods. Here he worked as a lecturer and private tutor. He was the first person to give talks on colour-blindness, based on his own affliction. For many years, colour-blindness was called 'Daltonism'. If you know someone who is colour-blind, it is probably a boy, since girls rarely suffer from it.

Dalton felt right at home at the Manchester Literary and Philosophical Society. Its active members became a kind of extended family for this shy man who never married. Manchester's 'Lit. & Phil.' was one of many similar societies established from the late eighteenth century in towns and cities throughout Europe and North America. Benjamin Franklin, the electrician, was one of the founders of the American Philosophical Society in Philadelphia. 'Natural philosophy' was, of course, what we now call 'science'. The 'Literary' in the Manchester society's name reminds us that science was not yet separated from other areas of intellectual activity; members would gather to hear talks on all sorts of subjects, from Shakespeare's plays to archaeology to chemistry. The age of specialisation, when chemists mostly talked to other chemists, or physicists just to other physicists, lay in the future. How exciting to range so broadly!

Dalton was a leading light in Manchester's scientific life, and his work was gradually appreciated throughout Europe and North America. He did some important experimental work in chemistry, but his reputation then and now rested on his idea of the chemical atom. Earlier chemists had shown that when chemicals react with each other, they do so in predictable ways. When hydrogen 'burns' in ordinary air (part of which is oxygen) the product is always water, and if you measure things carefully, you can see that the proportions of the two gases that combine to form water are always the same. (Don't try this at home, because hydrogen is very easily burned, and can explode.) This same kind of regularity also happened in other chemical experiments with gases, liquids and solids. Why?

For Lavoisier, in the previous century, this was because elements were the basic units of matter and simply couldn't be broken down into smaller parts. Dalton called the smallest unit of matter the 'atom'. He insisted that the atoms of one element are all the same, but different from the atoms of other elements. He thought of atoms as extremely small, solid bits of matter, surrounded by heat. The heat around the atom served to help him explain how his atoms, and the compounds they make when joined with other atoms, could exist in various states. For example, atoms of hydrogen and oxygen could exist as solid ice (when they had the least heat), or as liquid water, or as water vapour (when they had the most heat).

Dalton made models with little cut-outs to stand for his atoms. He marked his cardboard cut-outs with symbols, to save space (and time) when writing the names of compounds and their reactions (just as if he were sending a modern text message). At first his system was far too awkward to be used easily, but it was the right idea, so gradually chemists decided to use initials as the symbols for elements (and therefore Dalton's atoms). So hydrogen became 'H', oxygen 'O', and carbon 'C'. Another letter sometimes had to be added to avoid confusion: for example, when helium was discovered later, it couldn't be H so became 'He'.

The beauty of Dalton's atomic theory was that it allowed chemists to know things about these bits of matter that they could never actually see. If all the atoms in an element are the same, then they must weigh the same, so chemists could measure how much one weighed compared to another. In a compound made of different kinds of atoms, they could measure how much of each atom there was in the compound, by relative weight. (Dalton couldn't actually measure how much an individual atom weighed, so atomic weights were merely compared with the weights of other atoms.) Dalton led the way here, and he didn't always get it right. For instance, when oxygen and hydrogen combine to form water, he assumed that one atom of hydrogen and one atom of oxygen were involved. Based on his careful weighing, he gave the atomic weight of hydrogen as 1 (hydrogen was the lightest known element), and the atomic weight of oxygen as 7, so he said they had a weight ratio of 1 to 7, or 1:7. He always rounded his atomic weights to whole numbers and the comparative weights he was working with suggested he was right. In fact, the weight ratios in water are more like 1:8. We also now know that there are two atoms of hydrogen in each molecule of water, so the ratio of atomic weights is actually 1:16 – one of hydrogen to sixteen of oxygen. The current atomic weight of oxygen is 16. Hydrogen has retained the magical weight of 1, which Dalton gave it. Hydrogen is not only the lightest atom, it is also the most common one in the universe.

Dalton's atomic theory made sense of chemical reactions, by showing how elements or atoms combine in definite proportions. So, hydrogen and oxygen do this when they form water, and carbon and oxygen when they make carbon dioxide, and nitrogen and hydrogen when they make ammonium. Such regularity and consistency, as well as increasingly accurate tools for measurement, made chemistry a cutting-edge science in the early nineteenth century. Dalton's atomic theory provided its foundation.

Humphry Davy (1778–1829) was at the centre of this chemistry. Whereas Dalton was quiet, Davy was flamboyant and socially ambitious. Like Dalton, he came from a working-class background, and went to a good local school in Cornwall. He was lucky, too. He was apprenticed to a nearby doctor who was to train Davy to become a family doctor. Instead, Davy used the books that his master owned to educate himself in chemistry (and foreign languages). He moved to Bristol, becoming an assistant in a special medical institution that used different gases to treat patients. While there, Davy experimented with nitrous oxide – called 'laughing gas' because when you breathed it, it made you want to laugh. Davy's book on the gas, published in 1800, caused a sensation, for nitrous oxide had become a 'recreational drug' and nitrous oxide parties were all the rage. Davy also noted that, after breathing the gas, you didn't feel pain, and suggested that it might be useful in medicine. It took forty years before doctors took up his suggestion, and the gas is still sometimes used as an anaesthetic in modern dentistry and medicine.

Only the great city of London could satisfy Davy's ambitions. He got his chance to become lecturer in chemistry at the Royal Institution, an organisation that brought science to the middle-class public. Davy the showman thrived there. His talks on chemistry attracted large crowds – people often went to lectures for fun as well as to learn. Davy became a professor at the Institution, and his research flourished. Along with other chemists, he discovered the chemical use of Volta's electrical 'pile', the first battery. He dissolved compounds in liquids to make solutions and then used the pile to pass an electric current through them, analysing what happened. What he saw is that in many solutions, the elements and compounds were attracted to either the negative or the positive ends (poles) of the pile. Davy identified several new elements this way: sodium and potassium, for instance, which both accumulated around the negative pole. Sodium is part of the compound sodium chloride, the substance that makes the ocean salty, and which we put on our food. Once new elements were discovered, Davy could experiment with them, and work out their relative atomic weights.

Volta's pile, with its positive and negative poles, also changed the way chemists thought about atoms and chemical compounds. Positively charged things went towards the negative pole, and negatively charged ones to the positive pole. This helped explain why elements had natural tendencies to combine with each other. The Swedish chemist Jöns Jacob Berzelius (1779–1848) made this fact central to his famous theory of chemical combination. Berzelius survived a difficult childhood. Both his parents died when he was young and he was brought up by various relatives. But he grew up to become one of the most influential chemists in Europe. He discovered the joys of chemical research when he was training to be a doctor, and was able to work as a chemist in the Swedish capital, Stockholm, where he lived. He also travelled a lot, particularly to Paris and London – exciting places for a chemist.

Like Davy, Berzelius used the Voltaic pile to look at compounds in solution. He discovered several new elements this way, and he published lists of them with ever more accurate atomic weights. He worked out the weights by carefully analysing the relative weights of substances combining to make new compounds, or by breaking down compounds and then carefully measuring the products. His chemical table of 1818 listed the atomic weights of forty-five elements, with hydrogen still as 1. It also gave the known compositions of over 2,000 compounds. It was Berzelius who popularised Dalton's convention of identifying elements by the first one or two letters of their name: C for carbon, Ca for calcium, and so on. This made the language of chemical reactions much easier to read.

When compounds have more than one atom of an element in them, he indicated it with a number following the letter. Berzelius placed the number above the letter, but scientists now put it below: O_2 means there are two atoms of oxygen. Apart from that, Berzelius wrote chemical formulas much as we do today.

Berzelius was much better with inorganic compounds than with organic ones. 'Organic' compounds are ones containing carbon and are associated with living things: sugars and proteins are two examples. Organic compounds are often more complex chemically than inorganic ones, and they tend to react in rather different ways than the acids, salts and minerals that Berzelius was mostly examining. Berzelius thought that the reactions that go on in our bodies (or those of other living things such as trees and cows) could not be explained in quite the same way as those that happen in a laboratory. Organic chemistry was being developed during his lifetime in France and Germany, and although he distanced himself from these chemists, he actually contributed to their research. First, he provided the word 'protein' to describe one of the most important kinds of organic compounds. Second, he realised that many chemical reactions will not take place unless there is a third substance present. He called this third thing a 'catalyst'. It helped the reaction – often by speeding it up – but it did not actually change during the reaction, unlike the other chemicals that combined or broke down. Catalysts are found throughout nature, and trying to understand how they work has been the goal of many chemists since Berzelius's time.

Elsewhere in Europe, 'atoms' were helping chemists understand their work. There were still a lot of puzzles, however. In 1811, in Italy, the physicist Amedeo Avogadro (1776–1856) made a bold statement. It was so bold that it was neglected by chemists for almost forty years. He declared that the number of particles of any gas in a fixed volume and at the same temperature is always identical. 'Avogadro's hypothesis', as it came to be called, had important consequences. It meant that the molecular weights of gases could be calculated directly, using a formula he devised. His idea, or hypothesis, also helped modify Dalton's atomic theory, because it explained a curious feature of one of the most studied gases, water vapour. Chemists had long puzzled why the volume of hydrogen and oxygen in a particular amount of water vapour was incorrect if one assumed one atom of hydrogen and one of oxygen combined to make a molecule of water. It turned out that there were two atoms of hydrogen for every atom of oxygen in water vapour. Chemists discovered that many gases, including both hydrogen and oxygen, exist in nature not as single atoms but as molecules: two or more atoms joined together: H_2 and O_2, as we would say.

Avogadro's ideas didn't seem to make sense, if you believed Dalton's atomic theory, and Berzelius's idea of the atoms of elements having definite negative or positive characteristics. How could two negative oxygen atoms bind together? These problems meant that Avogadro's work was neglected for a long time. Much later on, though, it made sense of many chemical puzzles and is now fundamental to our understanding of the chemist's atom. Science is often like that: all the pieces only fit together after a long time and then things start to make sense.

CHAPTER 22

Forces, Fields and Magnetism

Dalton's atom helped create modern chemistry, but there were other ways of looking at atoms. For a start, they could do much more than just combine to make compounds. Atoms don't simply enter into chemical reactions. Both Davy and Berzelius had cleverly used the fact that atoms in a solution can be attracted to the positive or negative poles if an electric current is passed through the solution: atoms were part of 'electricity', too. In a solution of seawater, why would the sodium migrate to the negative pole, and the chlorine to the positive?

Such questions were hotly debated in the early nineteenth century. One of the chief investigators was Michael Faraday (1791–1867). Faraday was a quite remarkable man. Born into an ordinary family, he received only a basic education. He spent his youth learning bookbinding, but he discovered science and spent his spare time reading anything he could find about it. A popular children's book on chemistry fired his imagination, and a customer at the bookbinder's shop where he worked offered him a ticket to hear one of Humphry Davy's talks at the Royal Institution. Faraday listened in rapture and took careful notes in his neat handwriting. Ever keen, he showed his notes to Davy, who was impressed by their accuracy, but he advised Faraday that there were no jobs in science, and bookbinding was a better trade for a man who needed to make a living.

Shortly afterwards, however, a laboratory assistant at the Royal Institution was sacked, and Davy offered Faraday the job. He stayed there the rest of his life, helping to make it a profitable place with a great reputation. Faraday's early days at the Institution were spent in solving chemical problems for Davy. Faraday excelled in the laboratory, but he continued to read about more general scientific problems. He was a devout member of a particular group of Protestants; he devoted many hours to his Church, and his religious belief also guided his scientific enquiries. Quite simply, he thought that God had created the universe the way it is, but that human beings were capable of understanding how it all fits together.

Shortly after Faraday joined the Royal Institution, Davy and his new wife went on a tour of Europe, and they took Faraday with them. Davy's aristocratic wife treated Faraday as a servant, but the eighteen-month tour allowed Faraday to meet many of the leading scientific figures in Europe. Returning to London, Faraday and Davy continued to work on many practical problems: what caused explosions in mines; how the copper bottoms of ships could be improved; what were the optical characteristics of glass? As Davy became increasingly concerned with scientific politics, so Faraday became increasingly his own master, turning his attention to the relationship between electricity and magnetism.

In 1820, the Danish physicist Hans Christian Oersted (1777–1851) had discovered electromagnetism: the manipulation of an electric current so that it creates a magnetic 'field'. Magnetism had long been known, and the compass, with its iron needle always pointing north, is still useful. Navigators had used compasses long before Columbus discovered America, and natural philosophers had puzzled over why only a few substances (such as iron) could be magnetised. Most things could not. The fact that compasses always pointed in the same direction meant that the earth itself acted as a huge magnet.

Oersted's electromagnetism created a wave of scientific interest, and Faraday took up the challenge. In September 1821, he devised one of the most famous experiments in scientific history. Working with a small magnetic needle, he saw that the needle would continue to spin round if it was surrounded by wires carrying an electric current. While the electricity flowed through the coiled wire, it created a magnetic field to which the needle was continually attracted – spinning round and round. This was the result of what Faraday called 'lines of force', and he realised its significance. What he had done, for the first time, was to convert electrical energy (electricity) into mechanical energy (the movement or power of the rotating needle). He had invented the principle of all our electrical motors. These too convert electricity into power, in washing machines, CD players or vacuum cleaners.

Faraday continued to work with electricity and magnetism for the next thirty years. He was one of the most gifted experimenters who ever lived: thoughtful about planning his work and careful in carrying it out. His self-education had not included mathematics, so his scientific papers read much like his laboratory notebooks: detailed descriptions of his equipment, what he did and what he observed. His work also helped scientists understand the role of electrical charges in chemical reactions. By the early 1830s, he had added the electrical generator and the electrical transformer to his inventions. He made his electrical generator by moving a permanent magnet in and out of a coiled wire, which creates an electric current. To make his transformer, he passed an electric current through a wire wound around an iron ring, which caused a brief electrical current in another wire, wound around the opposite face of the ring. Faraday knew these experiments were crude, but he also knew he was on to something very important. The relationship between electricity and magnetism, and the conversion of electrical energy to mechanical energy, literally drive our modern world.

Faraday kept up his broad scientific interests, and spent much time sitting on scientific committees and running the Royal Institution. He started the Institution's Christmas Lectures, which are still hugely popular today – you may have seen one on the television. But electricity and magnetism remained his chief love. His fascination left us with a new vocabulary and many useful applications. He even made jokes about his inventions. When asked by a politician of the practical value of electricity, he is supposed to have said, 'Why, sir, there is every probability that you will soon be able to tax it!'

Across the Atlantic, another world-changing upshot of the great interest in electricity and magnetism appeared: the electric telegraph. Sending signals through electric wires started in the early 1800s, but the American Samuel Morse (1792–1872) developed the first long-distance telegraph. In 1844 he sent a message over thirty-eight miles (using the Morse Code that bears his name) from Washington, DC, to Baltimore. Telegraphic communication quickly developed all over the world, and the British used it to connect the outposts in their far-flung empire. It was now possible for people to communicate quickly with each other, and for news to be reported soon after it happened.

Faraday came up with the idea of a 'field' of action to explain why electricity and magnetism had their amazing properties. Fields (areas of influence) had been used by scientists before as they tried to explain the mysteries of chemical reactions, electricity, magnetism, light and gravity. These things took place, they thought, in a particular space or field, just as different games are played on a specific court, pitch or field. Faraday made this idea central to his explanation of electricity and magnetism, arguing that the important thing was to measure the area of activity rather than worry too much about what electricity, light or magnetism actually were. But the force of an electric field could be shown in experiments.

Faraday could not believe that something like gravity could exert its influence through a vacuum. Faraday solved this by assuming that there was no such thing as absolute emptiness.

Rather, he argued, space was filled with a very refined substance that was called the 'aether'. This aether (it's nothing to do with ether, the anaesthetic gas) made it possible for physicists and chemists to explain lots of things by direct influence. Thus, Faraday's 'fields' around electric currents or magnets could be the result of the current or magnet stimulating the very refined matter that constituted the aether. Gravity was easier to explain in this way, too: otherwise it seemed to be some strange occult force like the magical powers of the older alchemists, something that moderns like Faraday did not believe in. The aether was not something you could see, or feel, but physicists thought it explained the results of their experiments. In Britain, they continued to use the aether idea until the early 1900s, when experiments showed that it did not actually exist.

Much of Faraday's work on forces proved more useful. Later physicists extended it and provided better mathematical descriptions of electricity, magnetism and the many other phenomena that the physical world throws up when explored. Faraday was the last great physicist who didn't use mathematics.

The man who truly secured Faraday's legacy was James Clerk Maxwell (1831–79), one of the new breed of mathematical physicists. Maxwell is often spoken of in the same breath as Newton and Einstein. He was certainly one of the most creative physicists of all time. He was born in Edinburgh and educated there until he went to Cambridge University. He returned briefly to Scotland to teach, but in 1860 went to King's College, London. There he spent some of his most productive years. He had already described the rings of the planet Saturn, but in London he developed a theory of colour and took the first colour photograph. He was always interested in electricity and magnetism, and brought them firmly together: after Maxwell, physicists could use mathematics to describe electromagnetism. Maxwell provided the mathematical tools and equations to describe Faraday's ideas of the field. His equations showed that the electromagnetic force is a wave, and this was one of the most important discoveries in the whole of physics. This wave travels at the speed of light, and we now know that the light and energy from the sun come to us as electromagnetic waves. Indeed, Maxwell predicted the entire range of waves that we know: radio waves which allow radio broadcasts, microwaves in our kitchens, ultraviolet and infrared light waves above and below the colours of the rainbow, as well as X-rays and gamma waves, or rays. These waves are now part of everyday life. Yet most of these forms of energy were still to be discovered when Maxwell predicted them, so it was not surprising that it took some time for his genius to be appreciated. His *Treatise on Electricity and Magnetism* (1873) is probably the most important physics book between Newton's *Principia* and those of the twentieth century.

By the time he wrote this book, Maxwell had gone to Cambridge to organise the Cavendish Laboratory, where so much important physics research in the decades to follow would be done. Maxwell himself died young, aged forty-eight, but not before he had carried out fundamental research on how gases behave, using the special mathematical techniques of statistics. This allowed him to describe how the large numbers of atoms in a gas, each moving at slightly different speeds and in different directions, would produce the effects they do at different temperatures and pressures. He provided the mathematical tools to explain what Robert Boyle and Robert Hooke had observed all those years before. Maxwell also developed the basic concept of 'feedback mechanisms': processes that go in loops, which he called 'governors'. These mechanisms are very important in technology, in twentieth-century developments in artificial intelligence, and in computers. They also happen in our own bodies. For instance, when we get too hot, the body senses this and we sweat. The sweat cools our bodies as it evaporates. Or, if we are cold, we shiver, and the contractions of our muscles in the shiver produce heat which warms us. These feedback mechanisms help us maintain a constant body temperature.

Maxwell had a gentle sense of humour, was deeply religious and close to his wife, who kept a tight rein on him. At dinner parties, she was prone to say, 'James, you're beginning to enjoy yourself; it is time we go home.' Luckily, she didn't stop his pleasure in the laboratory.

CHAPTER 23

Digging Up Dinosaurs

When I was very young, I had problems telling the difference between dinosaurs and dragons. In pictures, they often look alike, with huge teeth, powerful jaws, scaly skin and evil eyes, and they are sometimes shown attacking some other creature around them. Both kinds of creatures are clearly the sort that it is best to avoid.

There is a significant difference between dinosaurs and dragons, however. Dragons appear in Greek myths, legends about England's King Arthur, Chinese New Year parades, and in many dramas throughout human history. But even if their power is such that they still feature in stories created today, they were always the products of the human imagination. Dragons never existed.

Dinosaurs, however, did once live. They were here for a very long time, even if human beings never saw them. They thrived around 200 million years ago, and we know about them because their bones have been preserved as fossils. The discovery of these bones in the early nineteenth century was an important step for science. First geologists, and then ordinary people, began to realise that the earth is far older than people had assumed.

The word 'palaeontology' was coined in France, in 1822, to give scientists a name for the study of fossils. Fossils are the outlines of parts of animals and plants that were once alive, but have slowly turned to stone (petrified) after they died, when the conditions are right. Fossils can be admired in many museums, and collecting them is fun. It is harder today, since a lot of the easy fossils have already been gathered for study and display. But in some places, like Lyme Regis on the south coast of England, the cliffs are still being eroded by the waves of the sea, and here fossils often come to light.

People have been coming upon fossils for thousands of years. Originally, the word 'fossil' just meant 'anything dug up', so 'fossils' might be old coins, pieces of pottery, or a nice quartz rock. But many of these objects buried in the earth looked like the shells, teeth or bones of animals, and gradually 'fossil' came to mean just these things that looked like bits of creatures. Shells of sea animals were sometimes found on mountain tops, far from the sea. Often the stony bones, teeth and shells didn't seem to be like those of any known animal. In the 1600s, when naturalists began to puzzle about what had been found, they developed three sorts of explanation. First, some believed that these shapes had been produced by a special force within nature, striving, but failing, to create new kinds of organisms. They were similar to living plants and animals, but hadn't quite made it. Secondly, others argued that fossils were really the remains of species of animals or plants that had simply not yet been discovered. So much of the earth itself remained unexplored, that these creatures would eventually be found in remote parts of the world, or in the oceans. A third group of scholars dared to suggest these organisms were creatures that had once been alive but were now extinct. If that was true, then the earth must be much older than most people thought.

It was not until the eighteenth century that the word 'fossil' got its modern meaning, that of the petrified remains of a plant or animal that had once been alive. The realisation of what this meant began to dominate scientific thinking. The scientist who convinced the world that some animals had become extinct was a Frenchman, Georges Cuvier (1769–1832). Cuvier was very good at anatomy, especially comparing the anatomy of different kinds of animals. He had a special interest in fish but also a vast knowledge of the whole animal kingdom. He dissected hundreds of different animals, then he compared the different parts of their bodies and explored what all their various organs did. He argued that animals are living machines in which every part has its proper purpose. He also noticed that everything in an animal's body worked together. For instance, animals that eat meat have canines (sharp teeth), which permit them to tear the flesh of their prey. They have the correct digestive system, muscles, and all other characteristics they need to catch and live on meat. Those that graze on plants, like cows and sheep, have teeth with flattened ends, which help in grinding grass and hay. Their bone structure and muscles are for standing around rather than running and pouncing.

Cuvier's belief that animals are so beautifully constructed that the whole fits together in harmony made it possible for him to say a lot about an animal's structure and mode of life just by looking at one part of it. Find a canine tooth and you have found a carnivore, he said, and he would apply the same principles to fossils. With another anatomist he undertook a thorough investigation of the fossils found around Paris. They discovered that the fossils often resembled parts in living animals that could still be found in the area, but in many cases the teeth and bones had small, but significant, differences. By chance, the frozen remains of a large elephant were found in Siberia. Cuvier examined this 'woolly mammoth', as it was called, and argued that it was not only unlike any known living elephant, but that an animal of this size would surely have been noticed before, were it still roaming around somewhere. So it must have become extinct.

When they accepted the idea that some species of animals (and plants) were now extinct, it was much easier for naturalists to interpret the large numbers of fossils that were then being uncovered. The discoveries of two rather unlikely people in England helped create the notion of a prehistoric world. The first of these was Mary

Anning (1799–1847). She was the daughter of a poor carpenter who lived in Lyme Regis, that place in southern England still being eroded by the sea. It was a brilliant place for Mary to hunt for fossils. Even as a young girl, she went fossil hunting, for good specimens could be sold to scientists and collectors. Mary and her brother Joseph used their local knowledge to develop a business collecting and selling fossils. In 1811 they found the skull, and then many of the other bones, of a strange creature. Estimated to have been seventeen feet long (five metres), it was unlike anything that had ever been found before. It was displayed in Oxford and was soon named *Ichthyosaurus*, which literally means 'fish-lizard', as it had had fins and so swam in water. Mary went on to find a number of other dramatic fossils, including one that had some resemblance to a giant turtle, but without any evidence that it had ever had a shell. This one was named *Plesiosaurus*, meaning 'nearly a reptile'. These discoveries brought her fame and some money. But as fossil hunting caught on, she found the competition fierce and had trouble supporting herself and her family through her business.

Mary Anning had little education and lost control over her fossil finds once she had sold them. Gideon Mantell (1790–1852) faced problems of a different kind. He was a family doctor in Lewes, Sussex – also in southern England – and had access to many fossils in limestone quarries nearby. As a doctor he had a good knowledge of anatomy and was able to interpret the fossils. But he had to fit his fossil work around a busy medical practice and a growing family. He turned his house into a kind of fossil museum, which didn't please his wife. Travelling to London to present his findings to the scientists there was a slow and expensive business.

Despite these problems, Mantell persisted, and was rewarded by uncovering several exotic beasts. In the 1820s, he found some teeth of a kind not seen before, and the original owner of the teeth was named *Iguanodon*, meaning 'having a tooth like an iguana' (a kind of tropical lizard). Some admirers gave him a more complete skeleton of the iguanodon that they had found. Mantell also discovered an armoured dinosaur, *Hylaeosaurus*, which confirmed that some of these gigantic creatures walked on land. Others were unearthed that had features of birds, so this strange world had creatures that lived in the sea, on the land and in the air.

When we see these enormous, wonderful creatures reconstructed in museums, it is difficult to understand how hard it was for the men and women who first uncovered them. The fossilised bones were often scattered and the skeletons had bits missing. They had only a limited number of living or fossilised animals to compare the findings with, and they had none of the modern techniques of dating their discoveries. They could only estimate the size of their finds by comparing the bones they had discovered – a thigh bone, for example – with large living animals, such as elephants or rhinoceroses. The estimated sizes were staggering. They used Cuvier's principle to help reconstruct whole skeletons from the parts and speculate on what the animal might have eaten, how it moved, and whether it lived on land, in the water, in the air, or some combination. A lot of their ideas had to be revised as more dinosaurs were discovered and more was learned about the early history of life on earth. But their findings changed forever how we think about the world we inhabit.

'Dinosaur hunters' made the general public realise how old the earth was, and how there had been complex creatures living long before human beings appeared. This ancient world captured their imagination, and fanciful pictures appeared in many popular magazines. Writers like Charles Dickens could refer to these giant reptiles, knowing that their readers would understand what they were talking about. The name 'dinosaur' was first used in 1842: it roughly means 'fearfully great lizard'. New kinds of dinosaurs continued to be uncovered, not only in England but elsewhere. They were quickly integrated into a general history of life on earth, and their period on earth was roughly calculated from the ages of the rocks in which they were found.

Richard Owen (1804–92), the man who gave them the name 'dinosaurs', used his own work on these creatures to further his scientific career. He was behind the building of what today is the Natural History Museum in London. It is a wonderful museum, and the dinosaurs still have a prominent place in it. Many of those on display are original specimens found by people like Mary Anning.

In 1851, London hosted the first in a series of World's Fairs. Called the Great Exhibition, it brought together displays of science, technology, art, transport and culture from all over the world. The Exhibition was housed in a building of amazing daring: the 'Crystal Palace', a gigantic glass house, located in the centre of Hyde Park, right in the heart of London. It was 33 metres tall, 124 metres wide and 563 metres long. People thought you couldn't build anything so large of glass and steel, but Joseph Paxton did. He was a gardener and builder who had experience constructing large greenhouses for Victorian gentlemen. The Exhibition was like nothing that had ever happened before, and six million people from all over the world flocked to see it during the six months it lasted.

When it closed, the Crystal Palace was taken down and moved to Sydenham Park on the southern edge of London. As part of the development of that site, the world's first theme park was created. It was devoted to dinosaurs and other creatures of the prehistoric world. Gigantic replicas of the *Iguanodon, Ichthyosaurus, Megalosaurus* and other beasts were constructed and placed in and around a man-made lake. The *Iguanodon* was so large that on New Year's Eve, 1853, twenty-four guests had dinner in the mould used to make its huge body. The area is still called Crystal Palace today, although the glass building burned down in a terrible fire in 1936. Some of the reconstructed dinosaurs don't look quite right now, but they survived the fire and can be seen today, battered and worn, but still magnificent reminders of the past.

We now know much more about the Age of Dinosaurs. Many different kinds have been found and we can date their ages more precisely than Mantell or Owen could. We sometimes say that they disappeared rather quickly. (Geological time is very slow, as we shall see in the next chapter.) What we mean to say is that the large dinosaurs went extinct, probably as a result of changes in the climate, after an enormous asteroid struck the earth about sixty-five million years ago. But not all of them disappeared. Some of the smaller dinosaurs survived and evolved, and you can see their descendants in your garden everyday. They are called birds.

CHAPTER 24

The History of Our Planet

Uncovering the bones of ancient beasts is only part of the story. Walking in the country, you must have noticed that a valley often has a river or stream running down the middle of it. Hills and mountains, too, will surround the valleys. In some parts of the world, say, the Alps of Switzerland, it is striking how the mountains are very high and the valleys are very deep.

How were the earth's features formed? Mountains and valleys could not have always been the way they are now, since the landscape is changed every year by earthquakes, volcanic eruptions, rivers and glaciers. The change in any one year may be slight, but even in your lifetime, visible differences occur. Coastlines wear away and houses sometimes fall into the sea. Multiply that by several, or many, generations, and the changes are even larger.

Violent earthquakes, volcanoes and tsunamis are nothing new. Mount Vesuvius, near Naples in Italy, erupted in AD 79. It buried the town beneath it, Pompeii, killing many people, and the volcanic ash and lava changed the coastline dramatically. Today you can walk along the streets of Pompeii, which have been excavated from the ash and pumice that settled there.

Many people wondered about what these kinds of dramatic happenings meant. Some thought they were supernatural acts. But from the late 1600s, observers began to study the earth as an object of natural history. Modern geology was born when they grappled with three problems. The first was a new way of understanding 'history'.

In earlier times, 'history' really meant 'description'. Natural history was simply a description of the earth and the things on it. Gradually, 'history' acquired its modern meaning of change through time. We are used to things changing quickly: clothes, music, hairstyles, slang, and anything to do with computers and mobile phones. We see photographs of people in the 1950s and think how different they looked then. This is not really new – the Romans dressed differently than the ancient Greeks, for instance – but the pace of change is much faster now. So, we accept change as natural. History is the study of that change.

The second problem was that of time. Aristotle assumed that the earth is eternal, and had always been much as it was when he lived. Ancient Chinese and Indian scientists also believed that the earth was very old. With the coming of Christian and Islamic views of the earth, time shrank. 'Time we may comprehend, 'tis but five days older than ourselves,' said the writer Sir Thomas Browne in 1642. What he meant was that the Book of Genesis tells the story of Creation, in which God created Adam and Eve on the sixth day. During the previous five days the earth, sky, stars, sun, moon and all the plants and animals were created. For Christians like Browne, our planet, the earth, was created only shortly before Adam and Eve saw the first dawn in the Garden of Eden.

If you read the Bible carefully, and add up all the ages of the descendants of Adam and Eve mentioned in the Old Testament, that gives an approximate date for the first couple. In the mid-1600s, an Irish archbishop did just that. His addition told him that the earth was created on 22 October 4004 BC, in the early evening, to be precise! Archbishop Ussher's calculations were not accepted by many Christians in the 1650s. But for people wanting to know how the geological features of the earth were formed, it was difficult to explain how, say, river valleys could have gradually come about if the earth was less than 6,000 years old.

This limited period of time also created difficulties explaining how shells could be found on mountain tops, far above the present oceans and seas. What geologists needed above all was to find more time for the earth to have been in existence. Then the things they were observing could be put into some kind of sensible perspective. And this they did. Beginning in the late seventeenth century, naturalists began to argue that the world *must* be older than the few thousand years allowed by Ussher. Several decades later, the Comte de Buffon (the pioneering natural historian we met in Chapter 19) worked out a scheme that combined cosmology and geology. His cosmology had the earth as originally a very hot ball, flung off from the sun. It gradually cooled down, and life became possible. He tentatively put the date of the separation of the earth from the sun at about 80,000 years ago, being careful with his exact language so as not to offend the Church.

The third problem was to understand the nature of rocks and minerals. All rocks are not the same. Some are hard, some soft and crumbly, and they are made up of different kinds of materials. They also seemed to be of different ages. Naming and analysing rocks and minerals allowed the geologists who studied them to put together a picture of the earth's history. Abraham Werner (1749–1817) in Germany did a lot of this early work. He worked in a university, but he was actively involved in mining. Mines deep under the earth helped scientists by providing samples of materials not easily obtained on the earth's surface. Werner based his classification of rocks not simply on their composition, but also on their relative ages. The oldest ones were very hard and never contained fossils.

Thus the kinds of rocks found in a given place provided a clue to the age of the place, relative to other places. Digging downwards, where the layers of rocks and earth (the *strata*, as geologists call them) contained fossils, these too provided clues to the relative ages of both the fossils and the strata in which they were found. The man who showed that the fossils were very important in this dating process was a surveyor, William Smith (1769–1839). Smith helped build Britain's canals in the early nineteenth century. Before railways, water was the best way to transport goods, particularly heavy things like coal. Smith measured many miles of land, helping to decide the best route for a new canal. What he gradually realised, as he created a geological map of England and Wales, was that the most important characteristic of a layer of the earth's crust was not simply the kind of rock it contained, but also the fossils that could be found within it.

With an expanded timescale for the earth's history, an understanding of the different kinds of rocks, and Smith's insight into the importance of the fossils, geologists could try to 'read' the earth's history. In the early 1800s, most geologists were 'catastrophists'. Piecing together the record uncovered through mining, canal building, and then railway building, they found many instances where volcanoes and earthquakes had thrown up layers previously buried deep in the earth's crust. So it seemed to most naturalists that the history of the earth had been one of periods of stability separated by periods of violent events – catastrophes – across the globe. Floods counted as catastrophes, so as geologists tried to fit their findings with the Bible, they were happy that there seemed to be evidence of massive and widespread flooding in the past, including a recent one (in geological terms) that could be the universal flood in which Noah took the animals two by two into his ark.

The catastrophists found a lot of evidence to support their view of the earth's history. The fossils in any of the various layers showed obvious differences from those below or above. The newer strata had fossils that were more like present-day living plants and animals than did the ones in the older layers. In Paris, Georges Cuvier (whom we met in the last chapter) was using 'comparative anatomy' and reconstructing vivid pictures of the animals of bygone ages. One of his followers was William Buckland (1784–1856), a liberal English clergyman who taught geology at Oxford University. Buckland was especially energetic in his search for geological evidence for the biblical flood. He found lots of things that he thought were obviously caused by water: debris washed into caves, and rocks and even huge boulders spread over the fields. In the 1820s, he was very sure that these were the result of Noah's flood; by the 1840s, as geological investigations had revealed more detail, he became less sure. Glaciers (huge rivers of ice) could have had an effect even in Britain, he realised. They provided a more convincing explanation of things like the scattered boulders, which could have been left behind as the ice moved slowly onwards.

In the 1820s and 1830s, most geologists believed that these ancient catastrophes coincided with new geological strata. Because the fossils in the layers were generally slightly different, they concluded that the earth's history consisted of a series of cataclysmic events – massive floods, violent earthquakes – followed by the creation of new plants and animals that were adapted to the new conditions that had come into being. The earth, it seemed, had undergone a progressive history in preparation for its crowning glory: the creation of mankind. This scheme fitted with the account of Creation in the Book of Genesis, either by assuming that its six days of creation were actually six long periods, or that the Bible only described the last creation, the age of human beings.

In 1830, Charles Lyell (1797–1875), a young lawyer-turned-geologist, challenged this general account. Lyell had looked at rocks and fossils in France and Italy. He was studying geology at Oxford and his teacher was William Buckland, the catastrophist. Lyell was dissatisfied with Buckland's geological vision. What could we show, Lyell asked, if we assumed that the geological forces that operated on the earth had actually always been uniform (the same)? He became the leader of the 'uniformitarians', who grew to be opposed to the 'catastrophists'. Lyell wanted to see how much of the whole geological history of the earth he could explain by using his principle of uniformity. He could see that at the present time the earth was very active geologically; there were still volcanoes, floods, erosion and earthquakes. If the rate of these changes was the same as long ago, was that enough to explain all the evidence of periods of ancient violent catastrophes? Yes, he said, and set out his reasons in a three-volume work, *The Principles of Geology* (1830–33). He would revise it for the next forty years, carefully taking into account his own and other geologists' research.

Lyell's uniformitarianism was a bold attempt to get rid of catastrophes and the reliance on miracles such as Noah's flood. He wanted to set geologists free to interpret the earth's history without interference from the Church. Lyell was a deeply religious man who held that mankind was a unique, moral creature, with a special position in the universe. And he saw more clearly than most that the catastrophists' idea of successive creations of plants and animals, approaching ever nearer those living in the present day, looked very much like evolution. Where the catastrophists compared deep fossils with shallow ones and saw progress, Lyell argued instead that fossils displayed no overall development. He was very excited when a fossil mammal was uncovered in an old layer, deep underground. Mammals were generally found only in recent strata, so this suggested to him that there was no real progress in the history of plants and animals, except for humans. If it looked like progress, that was only a fluke. No more than a tiny number of the species that existed in prehistoric times had been preserved as fossils.

Charles Lyell helped create modern geology. The way he thought about geology, and his extensive fieldwork, were both outstanding. He showed that, if our earth had a long enough history, much could be explained by simply observing what was going on now and using present-day geological events or forces to explain the past. A young naturalist, Charles Darwin, was much impressed with Lyell's *Principles of Geology*. He took the first volume with him (and had the other two sent out to him) when he set off on his travels around the globe on the *Beagle*. Darwin said that he looked with Lyell's eyes at the geological world – the world of earthquakes, rocks and fossils – during his voyage. But he came to very different conclusions about what the fossil record actually meant.

CHAPTER 25

The Greatest Show on Earth

Go for a walk in the countryside and you will find yourself among trees, flowers, mammals, birds and insects that belong in your part of the world. Go to a zoo and you will find exotic plants and animals from far away. Go to a natural history museum and there will be fossils, perhaps giant dinosaur skeletons, that are millions of years old. The person who taught us how all these living and fossil species are actually related was a quiet, modest man named Charles Darwin (1809–82). He changed the way we think about ourselves.

Carl Linnaeus (Chapter 19) named plants and animals with the idea that biological species are fixed. We still name them according to his principles. We can do this because, although we now know plants and animals do change, it's very slow. A biological *species* has real meaning. But with species there is *variation*. Children may differ from their parents: perhaps taller or with a different hair colour, or a bigger nose. Young fruit flies that swarm around rotting fruit in summer also differ from their parents, but because of their size it's hard to see. Easier to see are the differences between puppies or cats in a litter. What Darwin realised is that *variations* between parents and their offspring are very important, whether or not we see them. Even if we cannot always appreciate them, nature can, and does. Darwin's road to this vital insight was full of adventure and quiet thought.

Darwin's father and grandfather were successful doctors. His grandfather, Erasmus Darwin, had a theory of how plants and animals evolved, and wrote poems about science. Charles was a happy child, even though his mother died when he was eight. He discovered a love of nature and experimented with his chemistry set. He was only an average student at school. His father sent him to the University of Edinburgh to study medicine, but he was much more interested in natural history and biology. After the first surgical operation he saw made him physically sick, he knew he could never become a doctor. Darwin always remained extremely sensitive to suffering.

After his failure in Edinburgh, he went to the University of Cambridge to study for a basic arts degree, with the idea that he would become a clergyman. He passed his exams. Just. But Cambridge turned out to be vitally important because of the friendships he forged with the professors of botany and geology. They inspired him to become a naturalist. John Henslow took him plant collecting in the Cambridge countryside. Adam Sedgwick went with him to Wales to study the local rocks and fossils. After this tour with Sedgwick, Darwin had graduated from the university and was at a loose end, not sure what to do next. He was saved by an unusual offer: would he like to become the 'gentleman naturalist' on a surveying voyage aboard the ship HMS *Beagle*, led by Captain Robert Fitzroy of the Royal Navy? His father said no, but his uncle convinced his father that it was actually a great idea. The voyage on the *Beagle* was the making of Charles Darwin.

For almost five years, from December 1831 to October 1836, Darwin was away from home as the ship sailed gradually around the world. He was seasick for much of his time at sea, but he also spent plenty of time on land, especially in South America. He was an outstanding observer of all kinds of natural phenomena: landscapes, people and their customs, and plants, animals and fossils. He collected thousands of specimens and shipped them home, all carefully labelled. Today he would have written a blog, but he kept a wonderful journal, which he published after he came home. His *Journal of Researches* (1839) was immediately popular and remains a classic account of one of the most important scientific journeys ever taken. We know it as *The Voyage of the Beagle*.

Darwin's ideas about evolution would be worked out in the future, but even then he was privately wondering how plants and animals changed over time. His *Journal of Researches* told its readers about three especially important things. First, while Darwin was in Chile, he experienced – from the safety of the *Beagle* – a violent earthquake that dramatically raised the level of the coastline by almost fifteen feet (4.5 metres). Darwin had his copy of Lyell's *Principles of Geology* with him and was very impressed by Lyell's idea that violent events such as earthquakes could explain the past. The earthquake in Chile convinced Darwin that Lyell was right.

Second, Darwin was struck by the relationships between living species and recent fossils of plants and animals. On the eastern side of South America, he found large living armadillos, and fossils that were similar: *similar*, but clearly not of the same actual species. He discovered many other examples, and added his own to those found by other naturalists.

Third, and most famous, were his discoveries on the Galapagos Islands. This group of islands is separated by hundreds of miles from the western coast of South America. Here there were some amazing plants and animals, including giant tortoises and beautiful birds, many of which were unique to a single island. Darwin visited several of the islands and carefully collected specimens. He met an old man who could tell which island a turtle came from, so specific was the appearance of turtles from these islands. But it was only after Darwin returned to England that he began to realise the significance of what he had found. A bird expert looked at the finches collected from the different islands, and found that they were actually of different species. Each island of the Galapagos was, it seemed, a kind of mini-laboratory of change.

Leaving South America, the *Beagle* sailed across the Pacific to Australia, then under the southern tip of Africa. It returned to England via another brief visit to South America. When the ship arrived back in England in 1836, Darwin had become a first-class naturalist, very different from the nervous young man who had set out. He had also acquired a scientific reputation at home through the reports, letters and specimens he had sent back.

He spent the next few years working on many of the things he had collected on the expedition, writing three books. He also married his cousin Emma Wedgwood, and moved to a large house in the Kent countryside. Down House would be his home for the rest of his life, the place where he would do his most important work. It was just as well that he liked to be at home, since he suffered from a mysterious illness and he was often unwell. Whatever his illness was – and we still don't know what was wrong with him – he and Emma had nine children. He also wrote a steady stream of books and papers. One of these is the most important book in the whole history of biology: *On the Origin of Species*, published in 1859.

Years before that book was published, Darwin had begun keeping his private notebooks on 'transmutation'. He began the first in 1837, soon after he returned from the *Beagle* voyage. In 1838, Darwin read Thomas Malthus's *Essay on the Principle of Population*. Malthus, a clergyman, was mostly interested in why so many people are poor. He suggested that the poor marry too early and have more children than they can look after properly. Malthus said that all species of animals produce far more offspring than can survive. Cats can have three litters a year, each with six or more kittens. Each year an oak tree produces thousands of acorns, and each acorn can become another tree. Flies can produce millions of young flies each year. If all the offspring of these plants or animals survived, and if this happened in the following generations too, the world would soon be completely overrun with cats, oak trees or flies.

Malthus believed that all these extra offspring were essential because there is so much wastage. Nature is harsh – it's tough out there. When Darwin read Malthus's essay, he realised that he had discovered a reason why some young make it, and some don't. It would also explain why plants and animals change very gradually over long periods of time. Those that survive must have some advantage over their siblings, and there would be 'the survival of the fittest', or *natural selection* as Darwin called it. Darwin reasoned that all offspring inherit some traits from their parents, such as being fast runners. The offspring with the most useful traits were more likely to survive: they could run a bit faster, or had slightly spinier thorns. So those traits would be 'selected', because the less successful individuals, who did not have these traits, would not survive long enough to have offspring of their own.

Darwin realised that change in nature is very slow. But, he argued, we know that change can be much quicker when human beings are in charge of the process, selecting the traits they desire in their plants and animals. He called this *artificial selection*, and humans have been doing it for thousands of years. Darwin bred pigeons, and exchanged many letters with his fellow pigeon fanciers. He knew just how quickly the shapes and behaviour of their show pigeons could change, when the breeders carefully selected pigeons with certain traits for breeding chicks. Farmers had been doing the same thing with their cows, lambs and pigs. So had plant breeders when they tried to improve their crops, or produce more beautiful flowers. You know how very different a sheepdog is from a bulldog. It is easy to create variety in animals if the breeder selects the traits they desire.

Darwin saw that nature acts much more slowly, but, given enough time and the right environmental conditions, exactly the same thing happens. What he had learned of the birds and turtles in the Galapagos Islands illustrated how natural selection worked. The local conditions – soil, predators, food supplies – were slightly different on each island. So the local plants and animals had adapted to the differing local circumstances. The beaks of the various kinds of finches had been 'selected' for the different things they could find to eat: seeds, fruit, or ticks that lived on the tortoises. In some cases, as Darwin had learned, the differences had become great enough to create different species, although all the finches were still closely related. Time and isolation had allowed significant change to occur, and new species had evolved.

Silently, Darwin read widely and collected many other observations. He wrote a brief sketch of his theory in 1838 and a longer version in 1842. But he didn't publish his thoughts. Why? He wanted to be certain he was right. He knew he had a revolutionary view of the living world and that other scientists would criticise him severely if his account was not convincing. In 1844, Robert Chambers, an Edinburgh publisher and amateur naturalist, anonymously published his own version of species change. Chambers' *Vestiges of the Natural History of Creation* created a sensation. 'Transmutation' became a hot topic. Chambers had gathered a lot of evidence suggesting that living species are the descendants of previous ones. His ideas were rather vague, and he had no real theory about how this had happened. He made many mistakes. His book sold very well, but was savaged by the leading scientists – the very people Darwin hoped to convince. So Darwin waited. He finished some important publications from the *Beagle* work, and tackled an unusual but safe topic: barnacles. Dissecting and studying these small sea creatures was difficult, but Darwin always insisted that it gave him valuable insights into a group of animals with a large number of living and fossil species, each adapted differently to the way they lived.

After the barnacles, Darwin at last returned to his great work. In 1858, when he was writing a long book that he was calling 'Natural Selection', the postman delivered disastrous news. From far-away Asia came a letter asking for Darwin's opinion on a short paper. It was a brief account of the way natural selection could lead to species change over time. Darwin groaned. Its author, Alfred Russel Wallace (1823–1913), could have been summarising Darwin's own slow and painful path towards that same conclusion.

Darwin's friends Charles Lyell and Joseph Hooker, who both knew of his views on species, helped him out. They arranged for a joint presentation of Wallace's and Darwin's ideas at the Linnean Society in London. Nobody paid much attention to what was said at the meeting. Darwin was sick at home and Wallace didn't even know about it – he was 8,000 miles away. But Wallace's letter had persuaded Darwin that he must quickly write a summary of his ideas, instead of the long book he was working on. So *On the Origin of Species* was published on 24 November 1859. The publisher had 1,250 copies printed. They all sold in one day.

At the heart of his book were his two main ideas. First, natural selection *favours the survival of useful traits*, that is, characteristics that help individuals live and reproduce. (Artificial selection showed how human beings could dramatically alter the characteristics of plants and animals if they wanted to, illustrating how changeable plants and animals could be.) Second, natural selection, acting in the wild and over the long run, *produced new species*. They *evolved* slowly over time. The rest of the book was a brilliant demonstration of how well these ideas explained the natural world. Darwin wrote about the relationship between living species and their closely related fossil ancestors. He described the geographical distribution of plants and animals throughout the world. He explained how geographical isolation (as in the Galapagos Islands) provides the conditions in which new species can develop. And he emphasised that the embryos of some animals were surprisingly similar to the embryos of others. Darwin's *Origin* did for biology what Newton's *Principia* had done for physics. It made sense of a vast number of things in the natural world.

Darwin's biggest problem was inheritance: why offspring might be like their parents, and at the same time be slightly different from them and from each other. He read carefully and thought about it. He suggested some explanations, but he knew that heredity (genetics) was poorly understood, and he said so. He also knew that what was important was not saying *how* inheritance worked but that it *did happen*.

On the Origin of Species created a stir. People wrote and talked about it. Some had good things to say about it, others criticised it. Darwin simply kept working on it – he published six editions before he died. He developed his ideas, partly in response to criticisms, and partly because his own ideas continued to mature. As well as keeping the *Origin* up to date, he continued to write an astonishing number of other books on things that interested him: beautiful orchids, with their flowers adapted to the insects that pollinate them; plants that catch and digest insects; climbing plants that can cling to a wall; and even the humble earthworm. No wonder he was described as 'a man of enlarged curiosity'. Nothing seemed to escape his notice.

The *Origin* didn't say anything about human evolution, although Darwin knew that his insights were just as true for our own biological history. It was pretty clear to any reader of the first edition of the *Origin* that Darwin believed in the evolution of the human species, but he waited for more than a decade to say so openly, in *The Descent of Man* (1871).

Darwin made biological evolution a valid scientific theory. Some scientists were not convinced, but most were, even if they sometimes proposed their own versions of how and why it had happened. Many of the details of Darwin's great work have been corrected by later scientific work. It wasn't utterly perfect. It didn't have to be – science is like that. But from his study and garden at Down House, Darwin ensured that we would never look at life on earth in the same way again. The evolutionary history of our planet is simply the greatest show on earth.

CHAPTER 26

Little Boxes of Life

There are things we simply cannot see or hear. Many stars are beyond our gaze, and we can't see atoms, or even the tiny creatures that teem in puddles of rainwater. We can't hear sounds that many birds or mice can. But we can still learn about them, asking questions and using instruments that let us see or hear far better than with our eyes and ears alone. Just as telescopes let us see further into space, microscopes help us see further into the tiny building-blocks of living creatures.

In the seventeenth century, the pioneer of microbiology, Antonie van Leeuwenhoek, had used his small microscopes to look at blood cells and the hairs on a fly's legs. A century later, more advanced microscopes were allowing naturalists to examine these finer details of anatomy, and the wonderful array of tiny life. A 'compound' microscope could make things appear even bigger than a simple microscope. It is a tube with two lenses, the second of which magnifies the first image, so you get their combined magnification. Many thoughtful people did not completely trust microscopes. Early compound microscopes produced distortions or illusions of various kinds – for example, strange colours or lines where none existed. At the same time, there were only crude methods of cutting things into thin slices to examine them, and of trying to fix these slices on to a slide (a thin glass sheet). Consequently, many scientists thought using microscopes was not worth the effort.

Yet doctors and biologists wanted to understand how bodies work in the finest possible detail. In France, Xavier Bichat (1771–1802) began to investigate the different substances – what we call the 'tissues', whether hard like bone, soft like fat, or liquid like blood – that make up the human body. Bichat realised that the same kinds of tissues behaved in similar ways, no matter where they were in the human body. Thus, all muscles were composed of the same sort of tissue whether they were busy contracting in legs, arms, hands or feet. All tendons (the bits connecting muscle to bone), or the thin coating called serous tissue (like that surrounding the heart), were similar in all parts of the body. The study of cells and tissues is called 'histology' and Bichat was 'the father of histology'. Yet Bichat was one of those who were suspicious of microscopes, and he used only a simple magnifying glass.

Bichat's work inspired others to try to understand plants and animals in terms of their smaller, and more basic, building-blocks. In the early decades of the 1800s, there were several competing ideas about just what these fundamental building-blocks of plants and animals were. The technical problems of compound micro-scopes began to be solved in France and Britain from the late 1820s. From then on people looking down their instruments could be more confident that what they were seeing was an accurate picture of what was really there.

In the 1830s, the new microscopes helped two German scientists argue that the crucial building-blocks of life were cells, and that all plants and animals are composed of cells. One of these scientists was a botanist named Schleiden. The other was a doctor, Theodor Schwann (1810–82). Schwann explored how cells worked and how they were created. In the cells of plants and animals, activities take place that allow such things as movement, digestion, breathing and sensing. The cells act together, and they are the key to understanding how plants and animals function and live.

When you injure yourself – say, you cut your finger – more skin tissue will grow to heal the wound. But if tissues are made of cells, how are the new cells made? Schwann was very interested in chemistry, and he proposed that new cells crystallise out of a special kind of fluid, just as crystals can be grown in a laboratory from certain solutions. He wanted to explain how embryos develop in an egg, or the womb. He also wondered where the cells come from, those which appear if you get a scratch or a bruise. As a doctor, he could see that the area around an injury gets red and it may get full of pus cells. These pus cells, he thought, crystallise out of the watery fluid that we see as the swelling. It was an attractive theory, combining chemistry and biology, but it was quickly shown to be too simple.

As microscopes improved, more and more scientists began to watch what happens in cells. One of the most important cell-watchers was Rudolf Virchow (1821–1902). A man of wide interests, Virchow, mostly a pathologist, was also active in public health, politics, anthropology and archaeology. (He helped excavate the city of Troy, written about by Homer around 800 BC.) In the 1850s, Virchow began to think what the cell theory meant for medicine, and for the study of disease, known as pathology. Like Schwann, he saw cells as the basic units of living bodies. Understanding their functions in health and disease would be key to a new kind of medicine, based on science. He presented his ideas in a very important book called *Cellular Pathology* (1858). He showed that the diseases doctors see in their sick patients, and can later examine in the autopsy room (when studying their dead bodies), were always the result of events in the cells. These included the growth of cancer (which he was especially interested in), inflammation, with its pus and swelling, and heart disease. 'Learn to see microscopically,' he always taught his students in their pathology classes: peer down to the level of the cells.

Virchow combined his brilliant microscopic observations with a profound statement of a biological truth: 'All cells come from cells.' This is where he overtook Schwann. What he meant was that the pus cells in an angry swelling – after a splinter or a scrape, for example – actually came from other cells. They were not crystallised from body fluids. It also meant that cancer growths resulted from other cells, in this case cells that were behaving incorrectly and dividing when they should not. Every cell we can observe under the microscope has been produced by an existing cell (known as the 'mother' cell) dividing into two (the 'daughter' cells). Indeed, as biologists watched more and more, they sometimes saw this cell division taking place. And they noticed that the interior of the cells seemed to change when the cell divided into two. Something special was happening.

Earlier observations had already shown that the cell is not just a sack, full of the same kind of stuff. In the 1830s, an English botanist, Robert Brown (1773–1858), had argued that every cell has something at its centre: a *nucleus*, which is darker than the surrounding substance. Brown had looked at a lot of cells under his microscope and they all seemed to have this nucleus. The nucleus soon became accepted as a part of all cells. All the other material enclosed within the cell walls became known as *protoplasm*. This word means literally 'first mould', because at the time the protoplasm was viewed as the living stuff within the cells, whose functions gave life to plants and animals. In time other structures besides the nucleus were seen and named in cells.

Scientists quickly accepted the discovery of the nucleus and other parts of cells. But it was quite a different story for the very old debate about 'spontaneous generation', the observation that rotting meat and stagnant water seemed to spawn all kinds of tiny, but living, creatures. People knew that if they left a piece of uncovered meat on a table, in a couple of days they could expect to see maggots. They didn't know that flies lay eggs that hatch into maggots, so how could they explain where the maggots came from? Examine a drop of pond water under a microscope, and you will see that it is swarming with tiny creatures. How did they get there?

To nineteenth-century scientists, the easiest explanation was that these creatures had been made by, or generated from, their nourishing environments by a kind of chemical process. This was the common view, and it seemed to make sense. Since the maggots were not there when the meat was put down, how better to explain their presence than to assume that as the flesh decomposed it actually produced these rather disgusting creatures? Few people thought that complex things – elephants or oak trees – were spontaneously generated, but simple forms of life seemed to pop up without obvious explanation, except that they were somehow generated from their surroundings. Even Schwann's notion of living cells crystallising from the special bodily fluid was a kind of spontaneous generation, living cells coming from non-living material.

Naturalists in the 1600s and 1700s thought that they had shown that spontaneous generation does not occur, but the problem did not go away. It was hotly debated from the late 1850s by two French scientists. The winner finally convinced the scientific community that there was no spontaneous generation. But the story is not a simple one: the winner (who was correct) did not exactly play fair.

The first of these two French scientists was a chemist, Louis Pasteur (1822–95). In the 1850s he had begun to suspect that living cells could do quite extraordinary things. He was used to investigating the chemical properties of various compounds. He was also familiar with fermentation, the process in which grapes are mixed with yeast to make wine, and flour is mixed with yeast to make bread rise before baking it. Before Pasteur, fermentation was thought to be a particular kind of chemical reaction in which the yeast just acted as a catalyst – something to speed things up but remaining unchanged by the reaction. Pasteur would show instead that fermentation was a biological process caused by the yeast as it lived, feeding on the sugars in grapes and flour. The cells in the yeast were dividing to produce more cells, and in the process their living activities caused the desired alcohol in the wine or made the bread light and soft. Of course, these processes had to be stopped at the right time, by heating. If the yeast was allowed to go on and on living, the wine would turn to vinegar and the bread dough would eventually sink again. If this was happening in fermentation, Pasteur wondered about how other living micro-organisms might be involved in processes attributed to chemical reactions – such as spontaneous generation. So he turned it into a public competition with his fellow countryman, Félix Pouchet (1800–72), a supporter of spontaneous generation.

In a series of experiments, Pasteur boiled mixtures of straw and water to make them sterile. He then left them exposed to the air and the dust particles floating in it. Usually, if you examined the liquid after a few days, it would be teeming with micro-organisms. Pasteur showed that if you excluded the dust particles from the air, the solution would stay sterile. To show that these micro-organisms came with the dust particles, and not the air itself, he designed a special flask with a curved neck, like a swan's, that allowed in the air but not the dust. When Pouchet did similar experiments, his flasks contained micro-organisms after a few days. He interpreted his results as proving that spontaneous generation can occur. Pasteur assumed that when his experiments didn't turn out as he anticipated, it was because he hadn't cleaned his flasks well enough – and he presumed that Pouchet was always sloppy. Pasteur won the day, even if he quietly ignored the results of some of his experiments when they didn't give him what he wanted and appeared to support Pouchet! He triumphed partly because he was a dogged, determined scientist, who believed he was right, but also because Virchow's important statement that 'all cells come from cells' was gaining support. People wanted to believe Pasteur because his theories were a big step forward from old-fashioned ideas, and that's very important in science too.

Microscopy allowed great advances in medical and biological research. Microscopes were improved, and so were the tools to prepare specimens to examine under the lenses. Stains – special chemicals that acted like dyes – were especially important, because they could colour and highlight features of a cell's structure that would otherwise be overlooked. The stained nucleus, in particular, was observed to have a series of dark-staining strands that were given the name 'chromosomes'. (*Chromo* comes from the Greek for 'colour.') When a cell was dividing, the chromosomes could actually be seen to swell. The significance of this discovery, and of the other parts of the cell that scientists identified, had to wait until the twentieth century. But nineteenth-century doctors and biologists started the ball rolling. Above all, they showed that if you want to understand how whole plants and animals function, in both health and disease, you needed to start with the cells that they are made of. One kind of cell – single-celled organisms called *bacteria* – became especially important in understanding diseases. We are not done with Louis Pasteur, for he played a central role in the link between germs and disease, and in understanding how micro-organisms play their part in many aspects of our daily lives.

CHAPTER 27

Coughs, Sneezes and Diseases

If we have a runny nose, a cough or a stomach upset, we often say we have *caught* a bug or a virus, by which we mean some kind of germ. The notion of 'catching' something is so natural to us that it is hard to realise how amazing it was when someone came up with a theory that diseases can be caused by germs. Centuries before, doctors had explained that the ills people suffered were due to internal changes in the humours. Even more recently, doctors knew they could blame a bad constitution (we might say 'bad genes'), or too much indulgence in food or drink, or bad habits such as staying up all night. No one had thought that a living organism from the outside could cause a disease. It was a new idea, and it led to a major re-think about what disease itself actually means.

Doctors in earlier times had certainly talked about the 'seeds' of disease. The word 'virus' was often used too, but it then meant simply 'poison'. People dying from poison, accidental or deliberate, was nothing new. What was new with this theory of germs was that the external source was a tiny, living creature – a micro-organism. It brought with it a language of warfare: the body had 'defences' against this germ, and could 'fight' infection. Germ theory was a great turning point in medicine.

We met its most important champion, Louis Pasteur, in the last chapter. He came to germs gradually. He had been busy investigating the role of micro-organisms in many everyday events: the brewing of beer, the fermentation of wine, the baking of bread. The 'pasteurisation' of milk and other dairy products came to rely on discoveries he made: look in your refrigerator and you will see his name used. Pasteurised milk has been heated to just the right temperature, which kills the 'germs' in it. It will last longer and be safer to drink.

It was still a big step to show that bacteria, yeast, fungi and other micro-organisms could cause human and animal diseases. It was one thing to see these micro-organisms through a microscope, another thing to show that they and nothing else caused a particular disease. What we now call infectious diseases have always been killers. The bubonic plague, or Black Death, caused high fevers and very painful swellings on the body, known as buboes. It repeatedly swept through British towns and cities for more than 300 years from the 1340s onward. It was spread by fleas that lived on black rats, but moved on to humans when the rats died of plague, too. Smallpox, typhus, scarlet fever, with their skin rashes and high fevers, also took their grim toll. Parents might have eight or more babies and lose most of them to disease while they were still children.

When doctors studied these diseases, they explained them in one of two ways. Some thought these diseases of whole communities were *contagious*. That means they were spread from person to person by contact: when a healthy person touched a sick person or a sick person's clothes or sheets. Smallpox, with its horrible spots, seemed to be a contagious disease, especially since people who had not had the disease often came down with it if they nursed a friend or relative.

The spread of other diseases was much less easy to explain by contagion. Doctors had a theory that these diseases were caused by 'miasmas'. A miasma is a foul or unhealthy smell or vapour. Miasmatic diseases happened, they said, because of unhealthy disturbances in the atmosphere: the stench of rotting vegetation and sewage, the bad odours of the sickroom. During the 1800s, cholera was the most feared epidemic disease. It was common in India but in the 1820s it began to spread around the rest of the world. It took six years to travel from India to Britain, where it caused panic because it was a new and very frightening experience. Cholera caused dramatic diarrhoea and vomiting, leaving the poor victim shrivelled and in agony, dying an undignified death. It often killed in a day.

Today, international travel helps disease spread very quickly. In those days it made a slower progress. As European doctors and officials watched cholera spread slowly over Asia and Eastern Europe, they could not decide whether it was spreading from person to person (by contagion), or whether this was a miasmatic epidemic. Many people were worried that the disease was spreading through something everyone shared: the air they breathed.

Depending on which theory they believed, officials could do different things to try to stop disease spreading. If contagion was the cause, then it was best to isolate and quarantine the sufferers. For miasma, cleaning up and improving air quality were important. It was cholera that triggered the most intense debate when it first struck Britain in late 1831. In the panic, medical opinion was divided, but the quarantine measures did not seem to do much good. When the disease came again in 1848 and 1854, a London doctor, John Snow (1813–58), brilliantly worked out what was happening. By talking to local residents, and carefully mapping out each individual case in the neighbourhood, he became sure that the cholera was being spread by water from a public pump in Soho, central London. He believed it was contaminated with the faeces and vomit of cholera victims, and took a sample to examine by microscope. Although he could not identify any specific cause, his work emphasised that clean water was needed for public health.

Snow's research had shown how cholera was spread, not what caused it. For that, the laboratory was crucial, and especially the laboratory of Louis Pasteur. As he continued his research on micro-organisms, the French government asked him to investigate a silkworm disease that was destroying the French silk industry. Pasteur dutifully moved with his family to the south of France, where silk was being produced. He put his wife and children to work with him on trying to identify the cause of the problem. It turned out to be a micro-organism that was infecting the silkworm larvae. By showing how it could be avoided, Pasteur saved the French silk industry.

This put Pasteur on the disease trail. He wanted to demonstrate his belief that micro-organisms cause many of the diseases from which human beings and animals suffer. He began with anthrax, a disease of farm animals that was sometimes passed to humans. Until recently, this disease was largely forgotten, although it's one that terrorists now threaten us with. It causes nasty sores of the skin and, if it spreads to the bloodstream, it can kill. It is caused by a large bacterium, so it is relatively easy to detect. Anthrax was to become the first human disease Pasteur was able to prevent by making a vaccine.

Back in 1796, Edward Jenner (1749–1823), an English country doctor, had found a way to prevent smallpox by deliberately injecting a boy with cowpox, a similar but much milder disease. Cowpox was a disease of cows that milkmaids sometimes got, and it had been observed that these girls seemed to be protected from the more dangerous smallpox. Jenner called his new procedure *vaccination* (from the Latin word for cow, *vacca*), and vaccination programmes were started in many countries. They helped make this serious disease much less common.

Pasteur wanted to do something similar for anthrax, but there was no closely related disease around. Instead, he learned how to make the anthrax bacterium weaker, by changing its living condi-tions, such as the temperature, altering the food it could use, or exposing it to the air. Bacteria need the right conditions to flourish, just as we do. Pasteur succeeded in making his anthrax bacteria much less able to cause disease, and he called these weakened bacteria a *vaccine*, in honour of Jenner. Then he invited newspaper reporters to witness an experiment. Having injected some sheep and cattle with his vaccine, he gave the anthrax bacteria to that group, and to another. The experiment was an outstanding success: the animals he had vaccinated were unaffected when given the bacteria, whereas the unprotected animals died of the disease. Pasteur had made the world conscious of the power of medical science.

After anthrax came rabies. Rabies is a horrible disease, generally caused by a bite from an infected animal. It is often fatal, and its victims – including many young children – foam at the mouth and can't even drink water. The remarkable thing about Pasteur and rabies is that he could not even see what he was dealing with. The virus that causes rabies is so small that the microscopes available to Pasteur and his contemporaries could not bring it into focus. However, Pasteur knew from the victims' symptoms that whatever was causing rabies was attacking the brain and spinal cord, at the centre of the nervous system. So he used the spinal cord of rabbits to 'culture' (grow) the virus artificially. He could make it more, or less, harmful, according to the conditions under which he cultured it. He used his weaker virus to make a vaccine. His first human case was a dramatic success and made Pasteur a worldwide name. Joseph Meister was a young boy who had been bitten by a rabid dog. His desperate parents brought him to Pasteur, who agreed to try to save his life by a series of injections. Pasteur was a chemist, so a doctor actually had to give the injections, but the vaccination was a triumph. Young Meister survived, and worked for Pasteur for the rest of his life. Other people bitten by rabid animals hurried to Paris to receive this new miracle cure. The successful treatment created an international sensation, and people donated money to start a Pasteur Institute, where Pasteur worked until he died. The Institute is still going strong, more than a century later.

Pasteur was always unusual, both in his outstanding successes and in the ways he grew and studied his micro-organisms. Other scientists found his methods clumsy and difficult. Many of the laboratory tools that scientists still use to study bacteria were developed by Pasteur's German rival, Robert Koch (1843–1910).

Unlike Pasteur, Koch was a doctor, who began his work while treating patients. He, too, worked on anthrax, that bacterium that was easy to see. He worked out how anthrax moves from animals to humans and discovered that it has a complicated life-cycle. Sometimes the anthrax bacteria go into a kind of hibernation, known as the 'spore phase'. These spores are very hard to kill and they too can infect humans and animals so that they develop the disease in more than one way. Even though bacteria consist of only one cell, it turns out they are very complicated organisms.

Koch pioneered the use of photography to make a visible record of bacteria that cause disease. He grew his bacteria on a solid kind of jelly called agar-agar: this allowed individual 'colonies' (groups of bacteria) to be identified and studied. It was much less messy than Pasteur's flasks and soups. One of Koch's assistants, named Petri, invented the little dish used to hold the agar and grow the bacteria. Koch also appreciated the use of coloured stains to help identify different bacteria. These developments changed the face of bacteriology, and helped the international group of doctors and scientists begin to make sense of these tiny organisms.

Koch was a 'microbe hunter'. ('Microbe' is just short for microorganism.) He identified the germs that caused two of the most important diseases of the nineteenth century. In 1882, he announced his discovery of the tubercle bacillus, the bacterium that causes tuberculosis. Tuberculosis killed more people than any other disease in the nineteenth century, but doctors thought it was either inherited, or the result of an unhealthy lifestyle. Koch's research showed that tuberculosis is an infectious disease, spread from a sick individual to another person. It differed from other epidemic diseases such as influenza, measles, typhus and cholera, because it is a slow disease – slow to spread and infect, and slow to kill. Tuberculosis usually destroys the lungs over a number of years.

Koch's second great find was the bacterium that causes cholera, that other most feared disease. When it appeared in 1883 in Egypt, the French and the Germans sent scientists to see if they could uncover its cause. It was a kind of competition. One of the French team caught the disease and died. (Pasteur had wanted to go but was too frail.) Koch and his German colleagues thought they might have found the right germ, but they could not be sure. So Koch went on to India, where cholera was then always present. In identifying the cholera bacillus, he showed that Snow had been right – it was something in the water after all. He found the bacillus both in the diarrhoea of its victims, and in the wells from which they drew their water. Understanding the cause of infectious diseases paved the way for better control and, eventually, for vaccines, which have saved countless millions of lives over the past century.

From the late 1870s, many disease-causing germs were correctly identified (and many were announced that were later shown not to be dangerous at all). It was an exciting period, and a lot of doctors thought it heralded a new dawn for medicine and hygiene. It showed the importance of clean water, milk and everything else we eat and drink. From then on, doctors have advised us to wash our hands after using the toilet, and to cover our mouth when we cough. Identifying germs meant that scientists could make vaccines and, later, drugs. And it made modern surgery possible.

As early as the 1860s, the English surgeon Joseph Lister (1827–1912) had been inspired by Pasteur's germs. He introduced what he called *antiseptic* surgery. You probably have some antiseptic cream in your first aid kit. Lister's new method involved carbolic acid, also known as phenol, which was used to disinfect sewage. He would use carbolic acid to wash his surgical instruments and the bandages he would put over the body where it had been cut. He later invented a device to spray carbolic acid over the patient's body and the surgeon's hands during the operation. When Lister compared his patients with those of surgeons not using his 'Listerian' methods, or with his own pre-Listerian patients, he found that many more had survived their operation. They had not died from infections that started at the site of the operation and spread in the blood. In his experiments to disprove spontaneous generation, Pasteur had shown that 'germs' were carried through the air on particles of dust. Lister was killing these germs with his carbolic acid routine.

Just as he had improved on Pasteur's laboratory tools, so Robert Koch would advance Lister's antiseptic surgery. Lister had aimed to kill any disease-causing germs in the wound. Koch's *aseptic* surgery would prevent them getting into the wound in the first place. Koch invented the autoclave, a device that used very hot steam to sterilise surgical instruments. Aseptic surgery allowed surgeons to safely enter the body cavities (the chest, abdomen and brain) for the first time. It gradually brought about our modern operating theatre, with its surgical gowns and masks, rubber gloves and sterile equipment.

Along with modern hygiene, surgery could not have advanced without anaesthesia. It had been introduced into medicine in the 1840s, in America. Anaesthesia was a triumph for chemistry in the service of medicine, since the compounds that were shown to put people to sleep – ether and chloroform – were chemicals made in the laboratory. (Humphry Davy's nitrous oxide was another early anaesthetic.) The removal of agonising pain, and sometimes death, from surgery and childbirth seemed nothing short of miraculous. One of its British pioneers was John Snow, of cholera fame. Snow's anaesthetic career peaked when he gave anaesthesia to Queen Victoria, during the birth of her last two children. The Queen, who had had seven babies already without anaesthesia, thought it was a jolly good thing.

Understanding germs helped make advanced surgery possible. It also offered doctors ways to understand the infectious diseases which had caused so much pain and death throughout human history. There was now a scientific basis for Edward Jenner's discovery of vaccination to protect against specific diseases. These injections are worth it, even if they hurt at the time, for they offer hope that if everyone is vaccinated, many infectious diseases can be conquered. We know a lot more about germs than at the time of Pasteur and Koch. And we are more aware, as Chapter 36 will tell, how adaptable and slippery they are, these bacteria, viruses and parasites. They have been able to adapt to the medicines and treatments that doctors aim at them, and to become resistant – a lesson in Darwinian evolution. They survive because they adapt, a lesson that Darwin first taught.

CHAPTER 28

Engines and Energy

'I sell here, Sir, what all the world desires to have – POWER.' The engineer Matthew Boulton (1728–1809) knew what he was talking about. In the 1770s Boulton and other ambitious men, such as the inventor James Watt (1736–1819), were using steam engines in mining and manufacturing. They seemed to have tamed energy, or power. These men drove forward the Industrial Revolution in Britain, the first country to industrialise and to develop the factory system. It was a revolution driven by scientific advances, and relied on huge increases in power to manufacture goods at great speed and transport them far and wide. Our modern world is unimaginable without energy – lots of it. And it all started with steam.

Steam engines themselves are pretty simple. You can see the principle in action every time you boil a pan of water with a lid on: the force of the steam presses up on the lid to let the steam out and makes it rattle. Now imagine instead of a pan you have a closed cylinder with just a small hole in one end of it. Into this is fitted a moveable piston (that is, a disk that fits snugly into the cylinder, with a knob that fits snugly into the hole). The pressure of the escaping steam will force the piston up and move whatever might be attached to it: perhaps a rod with the wheels of a train attached to it. So a steam engine changes the energy of the steam into movement: mechanical energy. This engine can do useful work, such as driving a piece of machinery or pumping lots of water out of a mine.

Neither Boulton nor Watt invented the steam engine: they had been around for more than a hundred years. But the early models were crude, unreliable and inefficient. Watt, in particular, was the brain behind the improvement of the engine. His model not only provided the power that helped Britain industrialise, it also led scientists to investigate a basic law of nature. It helped them see that heat was not a substance, as Lavoisier had thought, but a form of energy.

Among the thoughtful people who were studying engines during the Industrial Revolution, one man in particular stands out from the crowd. This was a young French engineer, Sadi Carnot (1796–1832). The French and the British were great rivals at this time. Carnot was aware that the British had forged ahead in designing steam engines and using the power that they generated. He wanted France to catch up, and while watching steam engines do their work, he discovered a fundamental scientific principle. He was concerned with a steam engine's *efficiency*.

If a steam engine is perfectly efficient, it will turn to power *all* the energy needed to boil the water to drive the engine. You can measure the amount of heat produced by burning coal or wood to create the steam, then measure the power, or work the piston generated. If the engine were absolutely efficient, they would be exactly the same. Alas, absolutely efficient engines are impossible to build.

All engines have what is called a heat sump, or 'sink', where the cooled steam and water collect after doing their work. You can measure the temperature of the steam going in and the tempera-ture of the steam (or water) that is left at the end of each cycle. In the sump, the temperature is always lower coming out than it was going in. Carnot showed that you could use the difference between the two temperatures to calculate the efficiency of an engine. If perfect efficiency would score 1, then the actual efficiency is 1 minus the temperature in the sink (going out) divided by the temperature in the source (coming in). The only way to score the 1 of perfect efficiency would be to have the engine extracting *all* the heat out of the steam. Then, the ratio between out and in would be zero. That would give 1 - 0 = 1. For that to happen, one of the temperature measurements would have to be either zero or infinity: infinitely hot steam coming in or 'absolute zero' (the lowest temperature theoretically possible, which we will look at below) going out to the sink. Neither is possible, so efficiency is always less than perfect.

Carnot's simple equation, aimed at measuring the efficiency of engines, also summarises a deep law of nature. It explains why 'perpetual motion' machines are sometimes written about in science fiction, but can never exist in the real world. We always have to use energy to produce energy – for instance, we have to burn coal or some other fuel to heat the water in the first place. In the 1840s and 1850s, other scientists were working on this basic fact of nature. One of them was a German physicist, Rudolph Clausius (1822–88), who spent much of his life looking at how heat flows in carefully controlled experimental situations. To do this, he introduced a new concept in physics: *entropy*. Entropy is a measure of how mixed up (disordered) the things in a system are. It is much easier to mix things up than to unmix them. If you mix white with black paint, you get grey paint. The mixing is easy, but it's impossible to unmix them and get the pure black and white paints back again. If you stir milk and sugar in your tea, you can recover the sugar if you take a lot of trouble, but getting the milk back is impossible. Energy is no different: once you burn the coal, you can't use the heat it produced to get your coal back.

For people in the nineteenth century, entropy was a depressing idea. Clausius declared that the universe is becoming more and more mixed up, because entropy is its 'natural' stage. Once things get mixed up, it takes more energy to unmix them, just as it takes more energy to clean up a room than to get it messy. According to Clausius, the universe is slowly running down, and the end point will be a universe in which matter and energy are evenly distributed through all space. Even our sun will eventually die, in about five billion years, and with it, life on earth. In the meantime, of course, plants and animals, and human beings and our houses and computers, defy the ultimate endpoint of Clausius's insight. As the old saying has it, 'make hay while the sun shines'.

While physicists and engineers were worrying about the effects of entropy, they were also looking at what, exactly, energy was. Heat is an important form of energy, so the study of energy is called *thermodynamics* (a word that combines the Greek words for 'heat' and 'power'). In the 1840s several people came to similar conclusions about the relationships between different forms of energy. They were looking at a variety of things. What happens when water freezes or boils? How are our muscles able to lift weights? How do steam engines manage to use the hot water vapour to produce something than can do work? (The first public railway, driven by steam engines, had opened in the north of England in 1825.) Coming to the question from these different angles, they all realised that you cannot create energy out of nothing, nor can you make it completely disappear. All you can do with energy is to make it change from one form to another. Sometimes you can make this change do some work for you along the way. This became known as the principle of conservation of energy.

The Manchester physicist J.P. Joule (1818–89) wanted to understand the relationship between heat and work. How much energy does it take to do a certain amount of work? In a series of brilliant experiments, he showed that heat and work are directly related in ways that can be expressed mathematically. You use energy to produce work (to ride a bicycle, for instance), and heat is a common form of energy. Think about climbing to the top of a mountain. We use energy every time we move our muscles. This comes from the food that we eat and digest, using the oxygen we breathe to 'burn' the calories in our food. Now, there may be two paths to the mountain top, one very steep, and the other more gradual. What Joule showed is that, in terms of the energy needed, it doesn't matter which path you take. The steep path might leave you with aching muscles, but the amount of energy that you use in moving the weight of your body from the bottom to the top is the same, whichever path you take, or whether you run or walk up. Physicists still remember Joule's name. It is attached to several measurements, including a unit of energy, or heat.

People have long tried to measure how much heat an object contains, that is, its *temperature*. Galileo (Chapter 12) played around with a 'thermoscope', an instrument that changed as the temperature increased. A thermoscope allowed you to see that things were getting hotter or colder; a *thermometer* allowed you to put a number on the degree of heat. We still use two early attempts at devising a scale of temperatures. One was invented by the German physicist Daniel Gabriel Fahrenheit (1686–1736), who used thermometers containing both mercury and alcohol; in his scale, water freezes at 32 degrees, and our normal body temperature is 96 degrees. Anders Celsius (1701–44) devised his scale using the freezing and boiling points of water, with the former being set at zero degrees, and the latter at 100 degrees. His thermometer measured temperatures between these two points. These two scales are still part of our daily lives, from knowing what temperature to bake a cake at, to complaining about the weather.

The Scottish physicist William Thomson (1824–1907) invented another scale. He was especially interested in how heat and other forms of energy work in nature. He was a professor at the University of Glasgow and was later given the title Lord Kelvin. His temperature scale is known as the Kelvin or K scale. He worked out the K scale using very precise measurements and scientific principles. Compared with the K scale, Celsius and Fahrenheit turn out to be crude measures of temperature.

The K scale's defining point is the 'triple point of water'. This occurs when the three states of water – ice (a solid), water (a liquid) and water vapour (a gas) – are in 'thermodynamic equilibrium'. Thermodynamic equilibrium can happen in an experimental system, when a substance is insulated from its surroundings so that temperature and pressure are fixed. Then there is no change in the state of a substance and no energy escapes or enters the system. The triple point of water is when its solid, liquid and gas are held in perfect balance. As soon as the temperature or pressure changes then the balance or equilibrium is lost.

In Celsius and Fahrenheit, temperatures go into minus when it's very cold. You will have heard weather forecasters say 'minus two or three degrees'. There are no negative numbers in the K scale. Water freezes at 273.16 degrees Kelvin (as compared with 0 degrees in the Celsius scale and 32 degrees in the Fahrenheit scale). It gets a lot colder on the way down to 0 degrees Kelvin. But here 0 really means 0 or 'absolute zero'. At this impossibly cold temperature, all motion, all energy, stops. Just like the perfectly efficient engine, we cannot quite get there.

Kelvin and others helped to explain both the science and the practical workings of all kinds of engines. As the nineteenth century progressed, the three discoveries outlined in this chapter became the first, second and third law of thermodynamics: the conservation of energy, the 'law' of entropy, and the absolute still-ness of atoms at absolute zero. These laws help us understand important things about energy, work and power.

The modern world could not get enough of its new-found power: to run factories, ships, trains and – towards the end of Kelvin's life – motor cars. Trains and steam ships used the heat from coal in their furnaces to produce steam to drive the engines. But cars depended on a new kind of engine: the internal combustion engine. This needed a highly volatile fuel called petrol, or gasoline, which was discovered near the end of the nineteenth century. Petrol would become one of the most important products of the next century. Now, in the new millennium, it is still one of the most fought over and increasingly scarce resources in the world.

CHAPTER 29

Tabling the Elements

Every time we mix ingredients to bake something, we are using chemical reactions. The fizzing as we descale our kettles is chemistry at work for us. The plastic water bottles we carry, the coloured clothes we wear, are possible because of chemical knowledge gained over hundreds of years.

Chemistry became modern in the nineteenth century. Let's recap a little. At the beginning of the century, chemists embraced Dalton's original idea of the atom, as you read in Chapter 21. Then they made great strides in creating a special language that they would all understand, whatever country they came from. They had the system of symbols for elements, such as H_2 for two atoms of hydrogen. Everyone agreed that the *atom* was the smallest unit of matter. They used the word *element* for a substance made of only one kind of atom (carbon, for example). A *compound* was two or more elements bonded together chemically. You could break down compounds into elements (ammonium could be broken down into nitrogen and hydrogen), but once you got to the individual elements, you couldn't break things down any further.

Although atoms were clearly not the hard tiny balls that Dalton had suggested, it was extremely difficult to say exactly what they were. Instead, chemists began to discover a lot about how they acted when placed with other atoms or compounds. Some elements were simply not interested in reacting with others, no matter what you did. Some would react so violently together that you had to guard against an explosion. Sometimes, however, you could get a reaction if you helped to get it started. Oxygen and hydrogen could be placed together in a flask and nothing happened. If you put a spark to it, you had to watch out! Despite the dramatic explosion, the reaction produced nothing more unusual than water. At the other extreme, if magnesium and carbon were put together in a flask with no air, you could heat them forever and nothing would happen. Let in a bit of air, and you would be greeted with bright light and an awful lot of heat.

Chemists were aware of these various chemical reactions. They also became curious about what caused them and the patterns revealed in the laboratory. They set about their experiments in two main ways: synthesis and analysis. *Synthesis* is putting elements together: you start with single elements or simple compounds, and when these react with each other, you look at the results – at what has been made. *Analysis* is the reverse: you start with the more complex compound, do something to break it down, and, by looking at the end products, try to understand the compound that you started with. These methods began to give chemists a good idea of what many fairly simple compounds consisted of. It was then easier to create more complicated compounds, by adding new bits to substances they had a fair idea about.

From all these experiments, two things became particularly clear. First, as we have seen, the elements themselves each seemed to have either positive or negative tendencies. As the old saying goes, opposites attract. For example, sodium, a naturally positive element, combined easily with chlorine, a negative one, to make sodium chloride (which is just the table salt we sprinkle over food).

The positive and negative cancel each other out, so salt is neutral. All stable compounds (those that won't change unless something is done to them) are neutral even though they are made up of elements that were not necessarily so. Combining sodium and chlorine is an example of synthesis. You can do chemical analysis of the salt you've made. Dissolve the salt in water, put the solution in an electric field with its positive and negative poles, and it will split up. Sodium migrates to the negative pole, chlorine dances to the positive one. Hundreds of similar experiments convinced the chemists that the atoms of such elements do indeed have these positive and negative characteristics. And these characteristics play an essential role in determining what happens when elements react with each other.

Second, some groups of atoms may stick together during experiments, and these collective atoms can act like a single unit. These units were called 'radicals' and they too are positive or negative. They were especially important in 'organic' chemistry, where chemists were coming to understand a whole series of related compounds (all of which contained carbon), such as ethers, alcohols or benzenes. Benzenes were a fascinating group, each with a ring-like structure. Many chemists were eager to try to classify these organic groups, to understand what they were made of and how they reacted – not least because a lot of the substances were becoming valuable to industry. Increasingly, such industrial chemicals were made not in small laboratories, but in factories. The demand was growing for fertilisers, paints, medicines, dyes and, especially from the 1850s, oil products. The modern chemical industry had begun, and chemistry became a career, not just an indulgence for the curious or the rich.

The elements, too, have their own unique chemical and physical properties. As more and more were discovered, chemists found certain patterns. It appeared that individual atoms of some elements, such as hydrogen, sodium or chlorine, only wanted to combine with other atoms singly. For example, a single atom of hydrogen and one of chlorine combined to make a powerful acid, hydrochloric acid (HCl). A single atom of others, such as oxygen, barium and magnesium, seemed to have a double capacity to combine with other atoms or radicals, and so it takes two atoms of hydrogen to combine with oxygen to make water. Some elements were even more flexible, and there were always exceptions that made any hard and fast rules difficult to set down. Elements (and radicals) also differed in their eagerness to enter into chemical reactions. Phosphorous was so active that it had to be treated carefully; silicon was generally sluggish and much less dangerous.

The elements differed dramatically in their physical properties, too. At normal temperatures, hydrogen, oxygen, nitrogen and chlorine were gases; mercury and sodium were liquids. Most were naturally solids: metals like lead, copper, nickel and gold. Many other elements, above all carbon and sulphur, both intensively studied, were normally in a solid state. Put most solids in an ordinary furnace and they can easily be melted, and sometimes even vaporised (turned into a gas). Liquid mercury and sodium were also easy (if dangerous) to vaporise. Nineteenth-century chemists were not able to get low enough temperatures to turn gases like oxygen and nitrogen into liquids, much less solids. But they recognised their problem was a merely technical one. In principle, each element could exist in each of the three states of matter: solid, liquid and gas.

By the 1850s, chemistry was coming of age, and in this exciting period there was much to debate, about the relative weights of atoms, how molecules (groups of atoms) were bound together, the differences between 'organic' and 'inorganic' compounds, and much else. In 1860, something happened that helped create modern chemistry. It was something that today seems quite ordinary, but was unusual then: an international meeting. In the days before telephone, emails and easy travel, scientists rarely met and they communicated mostly by letters. Hearing another scientist from abroad talk about their work, with an open discussion afterwards, was a rare event. International meetings began in the 1850s, helped by more available travel by train and steamship, and they allowed people to meet and talk with their colleagues from other countries. They also announced to the world a belief widely shared by the scientific community: that science itself was objective and international, and above religion and politics, which often divided people and set whole nations at war with one another. The 1860 chemistry gathering met for three days in Karlsruhe, Germany. Many of the leading young chemists from all over Europe came there, including three who would direct chemistry for the rest of the century. The meeting's aims were set by the German August Kekulé (1829–96). He wanted chemists from different countries to agree on the words they should use to define the substances they worked with, and the nature of atoms and molecules. A fiery Sicilian chemist, Stanislao Cannizzaro (1826–1910), had already been arguing for this, and he gladly participated. So did a passionate Russian chemist, from Siberia, Dmitry Ivanovich Mendeleev (1834–1907). The delegates discussed Kekulé's suggestions for three days, and while no complete agreement was reached, the seeds had been sown.

At the meeting, copies of an article published by Cannizzaro in 1858 were given to many of the delegates. Here he reviewed the history of chemistry during the earlier part of the century. He called for chemists to take seriously the work of his fellow countryman Avogadro, who had clearly distinguished between an atom and a molecule. Cannizzaro also argued that it was vital to determine the relative atomic weights of the elements, and he showed how this could be done.

Mendeleev got the message. He owed much to his formidable mother, who had taken this last of her fourteen children from Siberia to St Petersburg, so that Mendeleev could learn about chemistry properly. Like many outstanding chemists of the time, Mendeleev wrote a textbook, based on his own experiments and what he taught his students. Like Cannizzaro, he wanted to bring order into the many elements that had been identified. Patterns had already been revealed: what were called the 'halogen' family – chlorine, bromine and iodine, for instance – reacted in similar ways. They could also be swapped for each other in chemical reactions. Some metals, such as copper and silver, also shared similarities in their reactions. Mendeleev began to list the elements in the order of their relative atomic weight (still using hydrogen as '1'). He presented his ideas in 1869.

Mendeleev did more than simply compile a list of the elements by atomic weight. He created a table, with rows and columns. You could read it across as well as up and down, and could see the relationship between elements with similar chemical properties. At first, his *periodic table*, as he came to call it, was very rough, and few chemists paid much attention to it. As he began to fill in the details, something interesting happened: there seemed to be occasional missing elements here and there, substances that his table implied should be there, but which had not been discovered. In fact, there was a whole missing column in his table, predicted by the relative atomic weights. Years later, this column turned out to be filled by non-reactive gases – called the 'noble' gases. Like aristocratic noblemen who don't mix socially with people below them, these gases were aloof from chemical reactions. The main ones were discovered only in the 1890s, and Mendeleev did not accept the findings at first. He soon realised that helium, neon and argon, with the atomic weights that they were shown to possess, had been predicted by his periodic table.

In the 1870s and 1880s, chemists discovered several more of the elements that Mendeleev had predicted on the basis of his table. Many chemists had dismissed as crazy speculation his predictions that the elements eventually called beryllium and gallium must exist. As the gaps he had identified gradually began to be filled, chemists appreciated the power of Mendeleev's table. It was guiding them to discover new elements in nature. It was also explaining what each element is like and how it reacts with other chemicals. What began as Mendeleev's attempt simply to understand the elements produced an amazing key to how nature behaves. The periodic table now hangs in classrooms and chemical laboratories all over the world.

For much of the nineteenth century, chemists had been concerned with chemical composition: which atoms and radicals made up specific compounds. The brains behind that first international chemical congress, August Kekulé, began to go further. He

encouraged scientists to aim to understand chemical *structure*.
Today's chemistry and molecular biology depend on scientists
knowing how atoms and molecules are arranged in a substance:
where they all sit, and the shapes they form. It would be impossible
to search for new drugs without this understanding, and Kekulé
pioneered it. He told of a dream in which he saw a chain of carbon
atoms curled around itself, like a snake biting its own tail. This
inspired one of his greatest insights, into benzene, the compound
of hydrogen and carbon, which has a closed ring structure. Radicals
or elements can be added at various points around the ring, and
this was an important advance for organic chemistry.

Dreams are one thing. Doing the hard slog is another. Kekulé
spent many hours in his laboratory, experimenting. He made sense
of organic chemistry – the chemistry of carbon compounds – and
taught the whole chemical world how to classify them in their
natural families. He was fascinated by carbon's flexibility in joining
with other chemicals. Methane gas, then widely used for heat and
light, was CH_4 – one carbon atom joined to four hydrogen atoms.
Two oxygen atoms could combine with a carbon atom, giving CO_2,
carbon dioxide. That these atomic preferences were not set in stone
was shown by the fact that carbon and oxygen could combine as
single atoms to create CO, the deadly gas carbon monoxide.

Chemists came up with a word for these joining patterns:
valence. And it could be deduced from the position of each element
in Mendeleev's periodic table. They speculated about why this was
so. Real understanding came only with the discovery, by the physicists, of the inner structure of the atom, and of the electron. The
electron linked the chemist's atom with the atom that the physicists
were studying, and the next chapter will tell that story.

CHAPTER 30

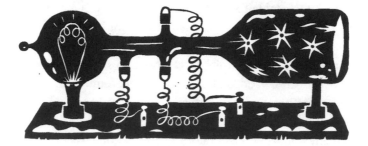

Into the Atom

The chemists liked the atom. It was what entered into chemical reactions. It had definite positions within compounds. It had properties that were roughly defined by its place on the periodic table. Each atom had its tendency to be either negative or positive in its relationships with other atoms, and to have the joining patterns called valence. Chemists also appreciated the difference between a single atom and the grouping of atoms into molecules (collections of atoms bound together). They realised that whereas most were happy to exist as single atoms, some atoms – hydrogen and oxygen, for instance – naturally existed in the molecular form (H_2 or O_2). Atoms' relative weights, with hydrogen always as 1, were also measured with increasing accuracy.

Yet none of this gave chemists much clue about the finer structures of atoms. They found they could manipulate atoms in their laboratories, but could not say much about what these units of matter actually were.

For much of the nineteenth century, physicists were rather more interested in other things: how energy was conserved, how electricity and magnetism could be measured, the nature of heat, and why gases behaved as they did. The physicists' theory of gases – called the kinetic theory – also involved thinking about atoms and molecules. But physicists, like chemists, agreed that although atomic theory was extremely useful in explaining what they saw and measured, the real nature of atoms was difficult to understand.

The first sign that atoms were not simply the smallest unit of matter came with the momentous discovery of one of its components, the *electron*. Experiments had already shown that atoms could possess electric charges, because electric currents in a solution attracted some atoms to the positive and others to the negative pole. Physicists were not so sure that an atom's electrical properties played any part in chemical reactions. But they measured their electrical charges and found they came in definite units. These units had been named 'electrons' in 1894, just after J.J. Thomson (1856–1940) in Cambridge began to use the cathode tube in his experimental work.

The cathode tube is quite simple. It is amazing, really, that something so simple could begin to tell us about the fundamental structure of the atom and the universe. This tube has had most of the air sucked out to create a partial vacuum, and electrodes have been inserted at each end. When an electric current is sent though the tube, all sorts of interesting things happen, including the production of rays (radiations). Radiations are streams of energy or particles, and those made in the cathode tube consisted mostly of fast-moving, charged particles. Thomson and his colleagues at the Cavendish Laboratories began to measure the electrical charge and the weight of some of these radiations. They tried to decide how these two measurements were related to each other. In 1897 Thomson proposed that these rays were streams of charged subatomic particles: bits of the atoms. He estimated that they weighed only a tiny fraction of the lightest atom, hydrogen. It took physicists several years to agree that Thomson had indeed found the electron – and that it was the unit of charge that he and others had been measuring for some time.

So, atoms have electrons. What else do they have? The answer came gradually, from the results of more experiments with the cathode tube. The vacuums within the tube became better, and stronger electric currents could be passed through. Among those who exploited these technical advances was Thomson's one-time student, collaborator and eventual successor at the Cavendish Laboratories in Cambridge, the New Zealander Ernest Rutherford (1873–1937). In the late 1890s Rutherford and Thomson identified two different kinds of rays given off by uranium, an element that had acquired great importance for physicists.

One of the uranium rays could be bent in a magnetic field; the other could not. Not knowing what they were, Rutherford called them simply 'alpha' and 'beta' rays – just 'A' and 'B' in Greek. The names stuck. Rutherford continued to experiment with both strange rays for decades. It turned out that not just uranium, but a whole group of elements, gave off (emitted) these rays. These elements created great excitement in the early years of the twentieth century, and they remain very important today. They are the 'radioactive' elements, and uranium, radium and thorium occur most commonly. When scientists began to investigate their special properties, they learned crucial things about atomic structure.

The alpha 'ray' was fundamental. (It is also called the alpha 'particle' – the distinctions sometimes blur in the very small and very fast world of atomic physics.) Rutherford and his colleagues aimed them at very thin sheets of metals, measuring what happened. Normally, the particles passed through the metal sheets. But occasionally, they bounced straight back. Imagine Rutherford's surprise when he considered what had happened. It was as if he had fired a heavy cannon-ball into a sheet of paper and discovered that it had bounced right back at him. What the experiment meant was that the alpha particle had encountered a very dense part of the atoms that were making up the metal sheet. This dense area was the *nucleus* of the atom. His experiments showed that atoms consist mostly of empty space, and that was why most of the alpha particles passed straight through. It was only when they hit the highly concentrated mass in this central nucleus that they bounced back.

More work showed that the nucleus is positively charged. Physicists began to suspect that the nucleus's positive charge is balanced by the electron's negative charges, and that the electrons circle around in the largely empty space surrounding the nucleus.

Rutherford is now considered the founder of nuclear physics. In 1908 he won the Nobel Prize in Chemistry for his discoveries. These prizes were named after their Swedish founder. They became the highest accolade in science after their introduction in 1901, and winning one remains the goal of many ambitious scientists. Rutherford was good at finding outstanding students and colleagues, and several of them won Nobel Prizes too.

Niels Bohr (1885–1962) from Denmark was one of these. He took Rutherford's idea that the atom's mass is almost all squeezed into its small nucleus and applied an exciting new tool called 'quantum' physics to develop something called the 'Bohr atom' in 1913. This was a model visualising what was going on inside the atom, using the best information scientists had at the time. It imagined that an atom had a structure something like our solar system, with the sun/nucleus in the middle and the planets/electrons spinning around it in their orbits. In Bohr's model, the weight of the positively charged nucleus gave the atom its atomic weight, and therefore its place in the periodic table. The nucleus was made up of positively-charged *protons*. The heavier the atom, the more protons were in the nucleus. The number of protons and electrons had to match so that the atom as a whole is electrically neutral. The electrons swirled around the nucleus in different orbits and this was where the 'quantum' bit came in. One of the brilliant parts of the whole package of ideas that scientists called 'quantum physics' was the idea that things in nature come in definite, individual packets ('quanta'). (The story of the quantum comes in Chapter 32.) The things can be mass, energy or whatever you are interested in. In the Bohr model, the electrons orbit in different, individual quantum states. The electrons nearest the nucleus are more strongly attracted to it. Those furthest away are the less strongly bound and it is these electrons that are available to participate in chemical reactions or to generate electricity or magnetism.

If this all seems rather difficult – well, it is. Bohr knew that. But he also knew that his Bohr model allowed physicists and chemists to talk in the same language. It was grounded on experiments by physicists, but also went far in explaining what the chemists observed in their own laboratories. In particular, it helped explain why elements in the periodic table behaved as they did, with their differing joining patterns, or valence. Those that joined singly did so because they had only one 'free' electron. Others had different patterns because of the number of 'free' electrons they had. His model of the atom has become one of the modern icons of science, even if we know now that the atom is much more complicated even than Bohr thought.

All sorts of new questions arose. First, how is it that the positively-charged protons can co-exist in the tight space that is the atomic nucleus? With electric charge, like repels like, and opposites attract (think of two magnets). So why don't the protons push each other apart, and why don't the electrons get sucked in? Second, the lightest known atom was hydrogen, so let's assume that hydrogen, with its atomic weight of 1, consists of a single proton and an almost weightless electron. That means it's reasonable to assume that the proton has an atomic weight of 1. So why don't the atomic weights of the atoms in the periodic table simply go up in a nice steady flow: 1, 2, 3, 4, 5, and so on?

An answer to the first puzzle had to wait until quantum mechanics was further developed. The second puzzle, about skips in the sequence of atomic weights, was solved much sooner, by another of Rutherford's Cambridge colleagues, James Chadwick (1891–1974). In 1932, Chadwick announced the results of his bombarding experiments. Ever since Rutherford, this method had been a vital tool for physicists at work on the structure of the atom. Chadwick had been sending streams of alpha particles at his favourite metal, beryllium. He found that beryllium sometimes emitted a particle with an atomic weight of one, and no charge. He used Rutherford's name for the particle – the *neutron* – but it soon became clear that it was not simply a proton and electron combined, as Rutherford had thought, but a fundamental particle of nature.

The neutron was a kind of missing link for physicists, explaining puzzling atomic weights and positions in the periodic table. Mendeleev's representation of the earth's elements was continuing to prove its worth in charting the basic materials of our planet. Chadwick's neutron also led to the discovery of *isotopes*. Sometimes atoms of the same element have different atomic weights – if they have a different number of neutrons, which are these neutral particles in the atom's nucleus. Isotopes are thus atoms of the same element with different atomic weights. Even hydrogen can occasionally have an atomic weight of 2 instead of 1, when it has a neutron along with its single proton. Chadwick won the Nobel Prize for his discovery of neutrons and what they could do, only three years after he discovered them.

The neutron was a powerful tool for bombarding the nuclei of other atoms. Lacking either a positive or negative charge, it isn't naturally repulsed by the heavily positive atomic nucleus, with its tightly packed protons. Chadwick recognised this, and saw that if you were going to smash atoms, you needed a machine that could accelerate them to high speeds and energies: a *cyclotron* or a *synchrotron*. These use very strong magnetic fields to propel atoms and their particles almost as fast as light. To do this kind of research, Chadwick left Cambridge for the University of Liverpool, because there he was given funds to build a cyclotron. There he saw that smashing high-speed neutrons into heavy atoms, such as uranium, could generate terrific energies. If such energies were harnessed, they could start a chain reaction leading to a momentous outcome: atomic 'fission', the splitting of the atom. The atomic bombs that were made and used to end the Second World War were the result of this work, and Chadwick was the leader of the British side of this project.

Many thought that Chadwick's discovery of the neutron solved the problems of the structures of atoms (the building-blocks of the universe). But they were wrong. There were many surprises still to be discovered. Yet even the basic understanding of the electron, proton and neutron had involved physicists with several waves and particles, such as the alpha, beta and gamma rays. They had had to understand other mysterious phenomena, such as X-rays, and the discovery that nature trades in those small packets, called quanta. Nuclear physics and quantum physics: these were the areas of physics at the cutting edge of knowledge for much of the twentieth century.

CHAPTER 31

Radioactivity

Have you ever broken a bone, or swallowed something by mistake? If so, the chances are you had an X-ray so a doctor could see inside your body without having to open it up. X-rays are routine today. At the end of the nineteenth century, they were a sensation. X-rays were the first kind of radiation to be harnessed, even before the meaning of radiation was properly understood. Radioactivity and atomic bombs came later.

In Germany, X-rays are still sometimes called 'Röntgen rays', after Wilhelm Röntgen (1845–1923). He was not the first to have seen their power, but he was the first to realise what he had seen. Science is often like that: it is not enough simply to see – you must understand what you are looking at.

In the 1890s, Röntgen, along with many other physicists (remember J.J. Thomson?) was working with the cathode ray tube. On 8 November 1895, he noticed that a photographic plate, some distance from his cathode ray tube, had mysteriously been exposed.

It was covered with black paper, and at that time scientists assumed that cathode rays had no effect that far away. He spent the next six weeks working out what was happening. Other scientists had observed the same thing but had not done anything about it. Röntgen discovered that these new rays went in a straight line, and were not affected by magnetic fields. Unlike light, they could not be reflected nor bent by a glass lens. But they could penetrate solid material, including his wife's hand! She posed for the first X-ray picture, her wedding ring clearly visible along with the bones of her fingers. Not knowing exactly what these rays were, he simply called them 'X-rays'. After the six weeks of hard work, he told the world.

X-rays became an immediate hit. Their medical uses were instantly recognised, in diagnosing broken bones or locating bullets or other things that shouldn't be lodged inside a body. Few things have ever been so instantly taken up by the general public. 'X-ray resistant' underwear was quickly for sale. Physicists debated what exactly X-rays were. After more than a decade of further research, X-rays were shown to be radiation with an unusually short wavelength and high energy. Early on, laboratory workers noticed that X-rays could damage human flesh, causing burns to appear, so they were used to try to kill cancer cells as early as 1896. It took a while longer for people to realise just how dangerous they were, and several of the early researchers died of radiation poisoning, or of a blood cancer called leukaemia. X-rays could *cause* as well as fight cancer.

While Röntgen worked with X-rays, another form of radiation – radioactivity – was discovered, this time in France. Henri Becquerel (1852–1908) was studying fluorescence, the way in which some substances glow, or naturally give off light. He was using a compound of uranium that did just that. When he discovered that this compound affected a photographic plate, just as Röntgen's X-rays had done, he assumed that he had discovered another source of this mysterious ray. But Becquerel found in 1896 that his rays did not behave like Röntgen's. They were a different kind of radiation, without the obvious dramatic effects of the X-rays that could 'see' through clothes or skin, but still worth another look.

In Paris, this challenge was taken up by the famous husband and wife physicists, Pierre and Marie Curie (1859–1906; 1867–1934). In 1898, the Curies obtained a tonne of pitchblende, crude tar-like stuff that contains some uranium. As they were extracting their relatively pure uranium, radioactivity burned their hands. They also discovered two new radioactive elements, which they named thorium and polonium, the latter after Marie's native Poland. As these elements had properties similar to uranium's, scientists around the world pressed to find out more about their powerful rays. These were the beta rays (streams of electrons); the alpha rays (shown in 1899 by Rutherford to be helium atoms with no electrons, and so positively charged); and gamma rays (without charge, but later shown to be electromagnetic radiation similar to X-rays). The Curies were truly heroic in their dedication to science. After Pierre was killed in a street accident, Marie continued their work, despite having their two young children to look after.

The ancient promise of alchemy, to see one element change into another, was almost fulfilled by the discovery of radioactivity. *Almost*, because the alchemists' dream had been to change lead or some other base metal into gold; what radioactivity did was transmute uranium into lead, a valuable metal into a base one! Still. Nature could do what the alchemists had merely dreamed of.

Like X-rays, radioactivity had important medical uses. Radium, another radioactive element discovered by Marie Curie, was especially valued. Its rays could kill cancer cells. But, like X-rays, radioactivity also causes cancer if the dose is too high. Many early workers, including Marie Curie, died from the effects of radiation, before proper safety guidelines were worked out. Her daughter, Irène, won her own Nobel Prize for work in the same field, and died early of the blood cancer that had killed her mother.

Uranium, thorium, polonium and radium are naturally radioactive. What does this mean? These radioactive elements are what physicists call 'heavy'. Their nucleus is very tightly packed and this makes it unstable. It is this instability that we detect as the radioactive rays. It was called 'radioactive decay' because when particles were lost, the element did literally decay, becoming a different element and taking up a different place in the periodic table. Studying this decay carefully continued that vital work of filling in the knowledge-gaps in the periodic table.

It also provided a valuable way of dating events in the earth's history, a process called 'radiometric dating'. Ernest Rutherford was a pioneer in this development, too, suggesting in 1905 that the technique would help in dating the age of the earth. Physicists calculated how long it would take for half of the atoms in a naturally radioactive element (uranium, for example) to decay away to its end product, the different version of the element (lead, in this example). This period of time was called the element's *half-life*. Elements' half-lives can vary from a few seconds to millions of years. Once they knew an element's half-life, scientists could date an event by looking in a fossil or a rock (any naturally occurring sample) to see how much there was of the original element and how much of the decayed one. The ratio between the two elements would tell them the age of the sample. One unusual form of carbon is naturally radioactive and its half-life can be used to date the fossilised remains of once-living animals and plants. All living things take up carbon through their lifetime. When they die this stops. So measuring the amount of radioactive carbon in fossils provides a date for their formation. Radiometric dating uses the same principle to date rocks, which gives a much longer time frame. The technique has transformed the study of fossils, because they are no longer just older or younger than each other – we know their approximate age.

Physicists quickly saw that enormous amounts of energy were involved in radioactive emissions. Naturally radioactive elements like uranium, and the radioactive forms of common elements like carbon, are scarce. But when you bombard atoms with alpha particles or neutrons, you can get many elements to artificially emit radioactivity's energy. This showed how much energy is packed into the atom's nucleus. Finding out how to make use of this potential has driven many physicists for the past hundred years.

When you bombard an atom and make it throw out an alpha particle from its nucleus, you 'split' the atom and make it a different element. This is nuclear *fission*. The nucleus has lost two protons. The alternative, nuclear *fusion*, occurs when an atom absorbs a particle and takes up a new place on the periodic table. Both fission and fusion release energy. The possibility of nuclear fusion was shown in the late 1930s by German and Austrian physicists, including Lise Meitner (1878–1968). Born a Jew, Meitner had converted to Christianity, but she still had to flee Nazi Germany in 1938. She discussed the fusion of two hydrogen atoms to form an atom of helium, the next element on the periodic table. From studying the sun and other stars, the conversion of hydrogen to helium was shown to be the main source of stellar energy. (Helium was discovered in the sun before it was found on earth: its atoms show characteristic wavelengths when examined with an instrument called the spectroscope.) This reaction needs very high temperatures, and in the 1930s it could not be done in the laboratory. But in theory, you could make a hydrogen bomb (a fusion bomb) that would release a vast amount of energy when it exploded.

In the 1930s, the alternative – the atomic or fission bomb – was more do-able. As the Nazis continued their aggression in Europe, war seemed increasingly likely. Scientists in several countries, including Germany, worked secretly towards preparing such devastating weapons. Crucial in this horrifying dance towards total war was the work of the Italian physicist Enrico Fermi (1901–54). Fermi and his group showed that bombarding atoms with 'slow' neutrons would cause the desired nuclear fission. Slow neutrons were passed through paraffin (or a similar substance) on the way to their target atom. At this reduced speed they were more likely to lodge in the nucleus, causing it to split. Fermi left Italy in 1938 to escape its Fascist regime, which was sympathetic to the Nazis. He went to the United States, as did so many of the most creative scientists (and writers, artists and thinkers) in the period. Today we sometimes speak of the 'brain drain', meaning that the best 'brains' leave their homes for better working conditions in other countries: more money, a bigger lab, a better chance to live their lives as they wish. People in the late 1930s and early 1940s fled because they had been sacked from their jobs and feared for their lives. The Nazis and Fascists did many horrific things. They also changed the face of science, and Britain and the United States gained most from this enforced brain drain.

In the USA, many of the refugees would join the top-secret 'Manhattan Project'. This was one of the most expensive scientific projects ever undertaken, but these were increasingly desperate times. By the late 1930s, the dramatic improvements in understanding the radioactive elements convinced many physicists that they could create a nuclear explosion. The difficulty was in controlling it. Some thought it would be too dangerous: the resulting chain reaction would simply blow up the whole planet. When war was declared in 1939, physicists in Britain and the USA believed that scientists in Germany and Japan would continue to work towards an atomic bomb and that the Allies must do the same. A number of scientists wrote to the American President, Franklin Roosevelt, urging him to authorise an Allied response. Among them was Albert Einstein, the world's most famous scientist and also a refugee from Nazi Germany.

Roosevelt agreed. At sites in Tennessee, Chicago and New Mexico, the many components of the fateful step were coordinated. The Manhattan Project was run along military lines. Scientists stopped publishing their findings. They put aside science's core value of openness and sharing information. War changes human values. The secret was not even shared with Communist Russia, a key ally of the USA and Britain, but still not trusted on the subject of top-secret bombs. By 1945, German, Japanese and Russian efforts to build atomic bombs had still not got very far, even though one of the scientists in the US secretly fed the Russians with information. But the Manhattan Project had produced two bombs. One used uranium, the other plutonium, a man-made radioactive element. A smaller test bomb was exploded in the American desert. It worked. The bombs were ready for use.

Germany surrendered on 8 May 1945, so no bomb was dropped in Europe. Japan continued its aggression in the Pacific. The new US President, Harry Truman, ordered the uranium bomb to be dropped on the Japanese city of Hiroshima on 6 August. It was detonated by firing one piece of uranium into another. The Japanese still did not surrender. Truman ordered the plutonium bomb to be dropped on a second Japanese city, Nagasaki, three days later. That action finally ended the war. The bombs had killed about 300,000 people, mostly civilians, and Japan surrendered. Everyone now saw the astounding power of nuclear energy. Our world was changed forever. Many of the scientists who had made these weapons of mass destruction knew their achievements had ended a terrible war, but worried about what they had created.

The incredible power of atomic energy continues to be important in our world. So, also, do its dangers. Mistrust between Russia and the USA continued after the Second World War, developing into the 'Cold War'. Both countries built up vast stores of atomic or nuclear weapons. Fortunately, they have not yet been used in anger, and although the stockpiles have been reduced over the years, through agreement, the number of nations that have nuclear weapons has grown.

The physics that was learned during the Manhattan Project has also been used to produce a more controlled release of energy. Nuclear power can generate electricity with only a fraction of the greenhouse gases released by burning coal and other fossil fuels. France generates almost three-quarters of its electricity by nuclear power, for instance. But the dangers of accidents and risks from terrorism have made many fearful of nuclear power, despite its benefits. Few things in modern science and technology better illustrate the mix of politics and social values than does the question: What should we do with our knowledge of nuclear energy?

CHAPTER 32

The Game-Changer
Einstein

Albert Einstein (1879–1955) is famous for his shock of white hair and his theories about matter, energy, space and time. And the equation $E = mc^2$. His ideas might be frighteningly hard to understand, but they changed the way we think about the universe. He was once asked what his laboratory looked like. By way of answer, he whipped out his fountain pen from his pocket. This was because Einstein was a thinker, not a doer. He worked at a desk or chalkboard rather than the laboratory bench.

Still, he needed the kind of information that could be gained by experiment, and in particular he would come to rely upon the work of the German physicist Max Planck (1858–1947). Planck was a thinker and an experimenter. He was about forty years old when he made his most fundamental discovery, at the University of Berlin. In the 1890s he started working on light bulbs, to see how he could produce a bulb that gave out the maximum light but used the least electricity. In his experiments he was using the idea of a 'black body', a hypothetical object that absorbs all the light shone onto it, reflecting none back. Think how hot you get wearing a black T-shirt in the sunshine, and how much cooler it is to wear a white one: the black clothing has absorbed the energy from the sunlight. So, the energy that comes with the light is absorbed by the black body. But it cannot simply store all this energy, so how does the black body give it back out again?

Planck knew that the amount of energy absorbed depends on the particular wavelength (the frequency) of the light. He took his very careful measurements of the energy and wavelength and put them into the mathematical equation $E = hv$. The energy (E) is equal to the frequency of the wavelength (v) multiplied by a fixed number (a 'constant' – h). In this equation the energy output Planck measured was always a whole number, not a fraction. This was important because being a fixed number meant that the energy came in individual little packets. He called each of these little packets a 'quantum', which just meant a quantity. He published his work in 1900, introducing the idea of the quantum to the new century. Physics, and how we understand our world, have never been quite the same since. The fixed number (h) was called 'Planck's constant' in his honour. His equation would prove just as important as Einstein's more famous $E = mc^2$.

It took some physicists a while to appreciate the real significance of Planck's experiments. Einstein was one who saw what it meant straight away. In 1905, he was working in the Zurich Patent Office as a clerk, and doing physics in his spare time. That year, he published three papers that made his name. The first, which won him a Nobel Prize in 1921, took Planck's work to a new level. Einstein thought more about Planck's black body radiation, and drew on the still-new quantum approach. After much thought, he showed – by some brilliant calculations – that the light was indeed being transmitted in small packages of energy. These packets moved independently of each other even though together they made up a wave. This was a startling claim, for physicists since Thomas Young a century before had analysed light in many experimental situations as if it were a continuous wave. It certainly generally behaved that way, and here was a young, still obscure, worker in a patent office saying that light could be a particle – a *photon*, or quantum of light.

Einstein's next paper from 1905 was equally revolutionary. This was where he introduced his Special Theory of Relativity, which showed that all movement is relative, that is, it can only be measured *in relation* to something else. It is a very complicated theory, but can be explained quite simply if you use your imagination. (Einstein was a great one for deep thinking about known data and exploring, in his mind, what would happen if . . .?) Imagine a train is moving out of a station. In the middle of one of the carriages there is a light bulb flashing on and off, sending out a flash at exactly the same time forwards and backwards, which is reflected in a mirror at each end of the carriage. If you were standing exactly in the middle of the carriage you would see the light bounce back from both mirrors at exactly the same time. But someone standing on the platform as the train went past would see the flashes one after the other. Although both flashes are still hitting the mirrors simultaneously, the train is moving forwards, so on the platform you would see the flash from the furthest-away mirror (at the front of the carriage) before you saw the flash from the closer mirror (at the back). So, although the speed of light remains the same, *when* it is seen is different depending on – or rather, relative to – whether the observer is moving or still. Einstein argued (with the help of some complicated equations, of course) that time is a fundamental dimension of reality. From now on physicists would need to think not simply of the three familiar dimensions of space – length, width and height – but of time, too.

Einstein showed that the speed of light is constant, no matter whether it is moving away from or towards us. (The speed of sound is different, which is why a train sounds different depending on whether we hear it approaching or moving away.) So the relativity in the Special Theory of Relativity doesn't apply to this constant speed of light. Instead the relativity occurs in the observers and in the fact that time needs to be included. Time is not absolute but relative. It changes the faster we travel and so do the clocks that record it for us. There is an old story about an astronaut travelling near the speed of light and coming back to earth to find that time has moved on. Everyone she knew has grown old and died. She is not much older that when she left, but as her clock has slowed down she is not aware of how long she has been away. (This is just a thought experiment and could happen only in science fiction.)

As if that wasn't enough, Einstein's famous equation $E = mc^2$ brought together mass (m) and energy (E) in a new way. The c is the velocity of light. In effect, he showed that mass and energy were two aspects of matter. Since the velocity of light is a very big number, and, when squared, an even bigger one, this means that only a very small amount of mass, if completely converted to energy, would be a lot of energy. Even atomic bombs convert only a tiny fraction of the mass to energy. If the mass in your body were totally converted into energy, it would have the force of fifteen large hydrogen bombs. Don't try the experiment, however.

Over the next few years, Einstein extended his thinking, and in 1916 he came up with a more general framework for the universe. This was his General Theory of Relativity. It introduced his ideas on the relationship between gravity and acceleration, and the structure of space. He showed that gravity and acceleration were actually equivalent. Imagine you are standing in a lift, and you drop an apple from your hand: it will fall to the floor of the lift. Now, if you let go of the apple at exactly the same moment that someone cuts the cord of the lift, you will fall along with the apple. It won't actually move, relative to you, as you both fall together. At any time, you can simply reach out and take hold of the apple. It will never reach the floor so long as the lift (and you) continue to fall. This is of course what happens in space, where there is no gravity. Astronauts and their spacecraft are essentially in free fall.

Einstein's General Theory of Relativity demonstrated that space, or rather space-time, is curved. It made predictions about several puzzling things that physicists had had difficulty in explaining. It suggested that light would be slightly bent when it passed near a large body. This was because light (made up of photons) has mass, and the larger body would exert a gravitational pull on the smaller mass of light. Measurements during an eclipse of the sun showed that this actually happens. Einstein's theory also explained curious features of the orbit of Mars around the sun, which Newton's less complex laws of gravity could not do.

Einstein had worked with the very small (the tiny photons of light) and the very large (the universe itself). He offered a compelling new way of bringing them together. In this he contributed to quantum theory as well as introducing his own ideas of relativity. These ideas, and the mathematics behind them, helped define the way physicists thought about both the large and the small. But Einstein did not approve of many of the new directions that physics was taking. He never lost his belief that the universe (with its atoms, electrons and other particles) is locked in a system of cause and effect. He famously said, 'God does not play at dice.' He meant that things always happen in regular, predictable patterns. Not everyone agreed, and other physicists who took on Planck's quantum ideas came to different conclusions.

The electron was central to much of the other early quantum work. Chapter 30 explained Niels Bohr's model of the quantum atom, in 1913. He had the electrons in fixed orbits with definite energies whizzing around the central nucleus. A lot of work was done trying to explain these relationships mathematically. Ordinary maths didn't work. To solve this problem, physicists turned to matrix mathematics. In ordinary maths, 2×3 is the same as 3×2. In matrix maths, this isn't so, and these special tools allowed an Austrian physicist, Erwin Schrödinger (1887–1961), to develop new equations in 1926. His wave equations described the behaviour of the electrons in the outer orbits of the atom. This was the beginning of quantum mechanics. It did for the very small what Newton had done for the very large. Like many of the physicists who changed the way we think about the world in the early twentieth century, Schrödinger had to flee the Nazis, and spent the war years in Dublin. Einstein, as we know, went to United States.

Schrödinger's wave equations brought some order into the picture. Then Werner Heisenberg (1901–76) came up with the 'uncertainty principle' in 1927. The principle was part philosophy, part experiment. Heisenberg said that the very act of experimenting with electrons changes them. This places limits on what we can know. We could know an electron's momentum (its mass multiplied by its speed), or its position, but not both. Measuring one affected the other. Einstein (among others) was appalled by this idea, and set about disproving Heisenberg's uncertainty principle. He couldn't. Einstein admitted defeat. So far the principle remains intact: there are simply limits to our knowledge of the very small.

The electron was also crucial to Paul Dirac (1902–84). This complex Englishman was considered to be almost another Einstein. His book on quantum mechanics led the field for three decades. His own equations about the quantum activities of atoms and sub-atomic particles were little short of brilliant. The trouble was, his equations required a strange particle – a positively charged electron – in order to work. This was like saying that there was both matter and anti-matter. The whole idea of 'anti-matter' was bizarre, since matter is the solid stuff of the universe. Within a few years, the search for such a particle was successful and the *positron* was discovered. This twin to the electron had a single positive charge. It combined with an electron, produced a burst of energy and then both particles disappeared. Matter and anti-matter could annihilate each other in less than the blink of an eye.

The positron showed physicists that atoms were composed of more than protons, electrons and neutrons. We look at some of these profound discoveries later, after physicists produced ever-higher energies to examine their atoms and particles. 'Examine' is not quite the right word. When working with high energy, physicists cannot actually see directly what is going on in their experiments. What they see instead are spots on a computer screen, or changes in the magnetism or energy of their experimental set-up. But atomic bombs, atomic energy and even the possibility of quantum computing all testify to nature's power and mystery – even if we cannot see it.

Max Planck's packet, or quantum, of energy, and Albert Einstein's realisation that mass and energy are merely two aspects of the same thing: these discoveries changed forever the way that the universe could be understood. Mass and energy; wave and particle; time and space: nature has revealed herself to be 'both . . . and . . .', not 'either . . . or . . .'. And while all this helped explain the structure of atoms and the creation of the universe, it also helps you get home at night. Satellites are so far above the earth that Satnav must include special relativity. If it weren't factored in, you could soon get lost.

CHAPTER 33

Moving Continents

Earthquakes are deadly and terrifying. Deadly because of the wholesale destruction they cause, terrifying because the earth should not move beneath our feet. And yet it does, all the time, if mostly unseen and unfelt. Like so much of science, understanding the earth's structure is about measuring the unseen, unfelt part – and convincing others that you're right. The continents and ocean floors *do* move beneath us.

What we experience of the earth's history in our lives is a tiny snapshot, the smallest of moments in a very long process. Geologists have scientific techniques, but they must also use their imaginations, thinking 'outside the box'. All good scientists do, even if they are working in the laboratory, checking their ideas against the evidence at hand.

Our nineteenth-century geologists used the traditional tools: fossil finds, analysing and classifying rocks, looking at the effects of earthquakes and volcanoes. All this they wove into a reasonable history of the earth. Much of what they learned still holds true today. But there were a number of problems that nagged at them, and needed a new kind of bold idea. The old 'catastrophists' had relied on the idea of different sorts of forces, or perhaps even miraculous interventions – great floods such as Noah's Flood described in the Bible. Instead, the new focus would be on time – immense periods of time called 'deep time'. What was the earth like, 200 million years ago, or twice or three times that number of years ago?

How could deep time help answer three key questions? First, why did the major continents look as if they could be cut out from the oceans and stuck together, like pieces in an enormous jigsaw puzzle? The east coast of South America would fit pretty snugly into the west coast of Africa. Was this an accident?

Second, why were the rock formations of South Africa so like ones found in Brazil, on the other side of the Atlantic Ocean? Why, in such a small island as Great Britain, were there dramatic variations between the Highlands of Scotland, with its crags and lochs, and the gently rolling Weald of Sussex in the south? Indeed, had Britain always been separated from the European mainland? Or Alaska from Asia?

Third, there were some odd patterns in the locations of plants and animals. Why were some species of snail found both in Europe and in eastern North America, but not on the other side of the American continent, on the west? Why were the marsupials in Australia so different from those found elsewhere? In the 1850s Darwin and Wallace pioneered some answers, and the theory of evolution helped explain a lot. Darwin ran some very smelly experiments, keeping seeds sitting in tubs of seawater in his study for months on end. He wanted to give the seeds an experience like a long sea journey. Then he planted them to see if they could germinate and grow. Sometimes they did, so that was one answer. Darwin also found ways to discover if birds could transport seeds, insects and other living things over very long distances. And they could, but this didn't explain all the puzzles.

There was one radical idea that could explain a great deal. This theory was that the continents had not always been where they are now, or that they had once been joined by strips of land, 'land bridges'. Many geologists from the late nineteenth century thought that there had once been land bridges in several places. There was good evidence that Britain had once been connected to Europe. It would explain very effectively why the fossil bones of bears, hyenas and other animals, not found in Britain in modern times, were found there. North America had once been connected to Asia across the Bering Strait, with animals and Native Americans undoubtedly crossing there. Land bridges joining Africa and South America seemed less likely, but the eminent Austrian geologist Eduard Suess (1831–1914) had a go at arguing this in his massive five-volume work (published between 1883 and 1909) on the earth. He said that the constant rising and falling of the surfaces of the earth during geological history made this possible. What was now sea-bed had once connected the two continents.

Not everyone was convinced, five volumes or not. Enter the German Alfred Wegener (1880–1930). Wegener was equally interested in the history of the earth's weather and its geology. In 1912 he gave a lecture on his theory of the continents moving: what would become 'continental drift'. The lecture became a book in 1915, and Wegener spent the rest of his life looking for further evidence. He died on the job, leading an expedition to Greenland to search for more clues to support his theory. Wegener's radical proposal was that, around 200 million years ago, there was only one large continent, Pangaea, surrounded by a vast ocean. This enormous continent had gradually broken up, with pieces of it literally floating on the ocean, like icebergs breaking away and floating on the sea. Unlike icebergs, which can melt and fade, the pieces of Pangaea became the new continents. And it wasn't over. Wegener thought the land masses were still moving apart, about ten metres a year. This estimate was way too high – recent measurements suggest a movement of only a few millimetres each year. But anything over a long enough period produces dramatic results.

Wegener had a few supporters, mainly in his native Germany, but most geologists found his ideas too far-fetched – too much like science fiction. Then, during the Second World War, submarines began the serious exploration of the ocean floor. After the war they revealed a new underwater landscape with enormous ridges of mountains and valleys, and extinct (and even active) volcanoes. Harry Hess (1906–69), a geologist working for the US Navy, traced these ridges and valleys and followed them on to the better-known dry land. He also followed the *fault lines*, those regions of the earth above and below water where earthquakes and volcanoes are common. What Hess discovered was that the land masses and the ocean floor were continuous, they ran into each other. The land didn't float as Wegener had suggested. How then might land masses move?

Hess was joined by physicists, meteorologists (weather watchers), oceanographers (studiers of the sea), seismologists (specialists in earthquakes) and the traditional geologists. They all began to try to work out the history of our earth, using the tools of these different sciences. This was not easy. The interior of the earth quickly gets very hot. Not that far down, instruments melt. So a lot of what we know about the composition and structure of our world's inner reaches had to be learned by indirect methods. Science is often like that.

Volcanoes spewing their molten lava had long been interpreted as the earth getting rid of the excess heat that had accumulated below, and in one sense, this is true. But it's not the whole picture. The discovery that radioactive elements, such as uranium, naturally release a lot of energy when they decay, added another source of interior heat. But radioactivity is an ongoing heat-producing source, and this meant that the older idea that the earth had once been a very hot ball but was now gradually cooling, was too simple.

At least, it was too simple for the geologist Arthur Holmes (1890–1965). He said that the earth gets rid of most of its continously generated internal heat by the familiar process of heat transfer, convection. The important bit was Holmes's realisation that it wasn't in the earth's upper crust – where we live – that things were happening, but in the next layer down towards the centre of the earth. This layer is called the mantle, and Holmes believed that the molten rocks there gradually move upwards, like the hotter water in your bath. As they move up and away from the hotter area, they cool, and sink down again, to be replaced by other molten rock, in a timeless cycle. It is some of this molten rock on the rise that spews out when volcanoes erupt. Most molten rock never makes it to the earth's surface, but spreads out as it cools and sinks, providing a mechanism to shift the continents apart, millimetre by millimetre.

As the depths of the oceans and earth were explored, a new way of working out the age of the planet added real meaning to deep time. The technique of radiometric dating had emerged from physicists' discovery of radioactivity (Chapter 31). Now it allowed the scientists to date the rocks they were studying by comparing the amounts of a radioactive element and its end product (uranium and lead, for example) in a rock sample. Using this technique, it was possible to know how old the rocks were, since after they are formed, no new material is incorporated in to them. Knowing the age of individual rock layers has in turn helped understand just how old the earth is. Rocks of more than four billion years have been found. Such old rocks are always on land. Those at the bottom of oceans are always newer. Oceans don't last as long as continents, and are in fact always dying and being reborn. This of course happens over a very long period of time, so don't worry about next summer at the beach. (On the other hand, man-made global warming may well keep melting the polar icecaps and lead to a dangerous rise in sea-levels in the coming decades.)

Rocks not only capture radioactive elements as they are formed, but also keep the magnetic orientation of their iron or other magnetically sensitive material. Like radioactivity, magnetism has helped earth scientists unravel the age of rocks. The earth's magnetic pole has not been constant over the long period of the earth's existence. North and south have flipped around on several occasions, so the north–south orientations can also provide evidence about when a rock was formed. Compasses will point north in our lifetimes and the lifetimes of our grandchildren, but things were not always so, and will not be so in the distant future, if the past is anything to go by.

Magnetism, convection, deep-sea landscapes and radiometric dating had revealed important clues about the ancient conditions of the earth. Taken together, they were enough to convince earth scientists that Wegener was almost right. Right, because continental movement did occur: sensitive measurements by satellites have confirmed the movement. But the drift or floating he suggested was wrong. Instead, John Wilson (1908–93) and others finished off the bold train of thought that Wegener had begun when they argued that the upper part of the earth's mantle is made up of a series of giant plates. These plates fit together, covering the earth, crossing the boundaries of land and sea. But they don't fit together perfectly, and it is at the joins that the fault lines appear. Understanding what goes on when one plate rubs against another, when they overlay each other or collide, is called *plate tectonics*. Think about the highest mountain on earth, Mount Everest in the Himalayas. Everest is as high as it is because the Himalayan Mountains were formed by two of these plates starting to collide with each other some seventy million years ago. There's no Nobel Prize in geology, but maybe there should be. Plate tectonics explains so much about earthquakes and tsunamis, mountains and rocks, fossils and living plants and animals. Our earth is a very old, but a very special place.

CHAPTER 34

What Do We Inherit?

Who do you look most like – mum or dad? Or perhaps a grand-father or aunt? If you are good at football or play the guitar or flute very well, does someone else in your family have these characteristics too? It has to be someone you are biologically related to and from whom you could have inherited these things, not just a relative by marriage, like a stepmother or stepfather. These relatives can do wonderful things for you, but you cannot inherit any of their genes.

We know now that things like the colour of our eyes or hair are controlled and passed from one generation to the next through our genes. *Genetics* is the study of our genes. *Heredity* or *inheritance* are the words we use to describe how the information our genes possess is passed on. Our genes determine an awful lot about who we are. So how did people realise these tiny things were so important?

Let's go back to Charles Darwin for a moment (Chapter 25). Heredity was central to Darwin's work. It was vital to his ideas on the evolution of species, even if he hadn't worked out how heredity occurs. Biologists continued to debate how it happens long after his book *On the Origin of Species* was published in 1859. In particular, they were interested in whether 'soft' heredity can sometimes happen. Soft heredity was an idea associated with a French naturalist, Jean-Baptiste Lamarck (1744–1829), who also believed in the development of species by evolutionary change. Think about a giraffe's long neck: how had that evolved over time? Lamarck said it was because as giraffes continually stretch upwards to reach the leaves on the tallest trees so this slight change will be passed on to their offspring generation after generation. Given enough time and enough stretching, a shorter-necked animal would eventually become a longer-necked one. The environment would interact with the organism, shaping or adapting it, and that would be passed on to the following generations.

Trying to prove soft heredity experimentally was very difficult. Darwin's cousin, Francis Galton (1822–1911), performed a careful series of experiments, in which he introduced the blood of black rabbits into white ones. The offspring of the transfused rabbits showed no sign of being affected by the blood. He cut off rats' tails for generations on end, but did not produce a race of tailless rats. Circumcising young boys had not had any effect on future generations of male babies.

Arguments for and against were bandied about until the early 1900s. Then two things convinced most biologists that the traits plants or animals had simply acquired during their life are not passed on to their offspring. First came the rediscovery of the work of a monk from Moravia (now part of the Czech Republic), Gregor Mendel (1822–84). In the 1860s, Mendel had published (in a little-read journal) the results of his experiments in the monastery garden. He had become fascinated with peas, even before Galton was cutting off the tails of his rats. Mendel wondered what happened when pea plants with certain characteristics were carefully 'crossed' (that is, plants with differently coloured peas were bred together), to provide the next generation of pea plants. Peas were good to work with because they grew fast, so it was quick and easy to move from one generation to the next. And, in the pod they also had clear differences – the peas were either yellow or green, in wrinkled or smooth skins. He discovered that these traits were inherited with mathematical precision, but in ways that could be easily overlooked. If a plant with green peas (its seeds) was crossed with a yellow one, all the first generation of peas were yellow. But when he crossed these first-generation plants with each other, in the second generation three of every four plants would have yellow peas and one would have green. The yellow trait had dominated in the first generation, but in the second, the 'recessive' trait (the green) showed itself again. What did these strong patterns mean? Mendel concluded that heredity is 'particulate', that is, that plants and animals inherit traits in separate units. Rather than the little-by-little changes of soft heredity, or some average of the attributes of the two parents, heredity was something quite definite. Peas were either green or yellow, and not some shade in between.

While Mendel's work lay unnoticed, August Weismann (1834–1914) provided the second critical assault on soft heredity. Where Mendel was mostly concerned with his religious life, Weismann was first and foremost a determined scientist. A brilliant German biologist, he strongly believed that Darwin's evolutionary views were correct. But he could see that the lack of a good explanation for heredity was a problem. He turned his own fascination with cells and cell division into a solution.

A few years before Mendel's experiments with his peas, Rudolf Virchow had announced his ideas about cell division (Chapter 26). In the 1880s and 1890s Weismann saw that to make an egg or a sperm cell, 'mother' cells of the reproductive system divided in a way that was different from cell division in the rest of the body. It was this difference that was the key. Known as the process of *meiosis*, here the chromosomes divided and half of the chromo-somal material went into each of the resulting 'daughter' cells. In all the other body cells, the 'daughter' cell has the same amount of the chromosome material as the 'mother'. (If you're confused, remember that a 'mother' cell is just any existing cell and that it splits into two 'daughter' cells. They are found throughout the body and have nothing to do with real mothers and daughters.) So when the egg and sperm cells fused, the two halves of the chromosomal material would make up the full amount again in the fertilized egg. These reproductive cells were different to all the other cells of the body. Weissman argued that it did not matter what else happened to the cells of the muscles or bones or blood vessels or nerves: only these reproductive cells contained what would be inherited by the individual's offspring. So in the case of the giraffe's neck the supposed stretching would have no effect on the egg and sperm cells, and it was these cells that contained what he termed the 'germ plasm'. It was the germ plasm, on the chromosomes of the egg and sperm cells, which was inherited, and he called his idea of heredity the 'continuity of the germ plasm'.

In 1900 not one but three separate scientists dusted off copies of the journal with Mendel's article in it. They alerted the scientific world to the results of Mendel's pea experiments. Biologists realised that Mendel had provided the best experimental evidence yet for Weismann's 'continuity of the germ plasm' and that 'Mendelism', as it was soon called, had a sound scientific basis.

The scientific community was soon split into two groups, the 'Mendelians' and the 'biometricians'. The biometricians, led by the statistics expert Karl Pearson (1857–1936), believed in 'continuous' inheritance. They thought that what we inherit is an average of the attributes of our parents. They conducted important fieldwork in measuring very small differences in sea creatures and snails. They showed that such small differences could play a significant role in determining how many offspring survived – what is termed the reproductive success of species. The Mendelians were led by the Cambridge biologist William Bateson (1861–1926). He coined the term 'genetics'. Mendelians emphasised the inheritance of the sort of discrete (separate) traits that the monk had illustrated. They argued that biological change occurred by leaps, rather than the slow, continuous changes of the biometricians. Both groups accepted the *fact* of evolution: they merely argued about how it happened.

These arguments were fierce for about twenty years. Then, in the 1920s, several people showed that each group was right and wrong at the same time. They were just looking at two different sides of the same problem. Many biological characteristics are inherited in a 'blending', 'biometrical' fashion. A tall father and a short mother will have offspring that average out or 'blend' their heights. Some of the children may be as tall as the father (or even taller), but the average height will tend to be midway between the two parents. Other characteristics, such as human eye colour, or the colour of peas, are inherited in an either/or, not a both/and fashion. The differences between the Mendelians and the biometricians were resolved when they measured whole populations, and then appled mathematical reasoning to the problem. These new biologists, such as J.B.S. Haldane (1892–1964), appreciated the brilliance of Darwin's original insights. They realised that in any population there is random variation that can be inherited. If it gives an advantage, those plants and animals that have it will survive and other kinds of variation will die out.

How we inherit what we do is also vitally important too. This was the next part of the puzzle. Much of the early work was carried out in the laboratory of Thomas Hunt Morgan (1866–1945), at Columbia University in New York City. He began his career looking at how animals begin life and develop as embryos. He never completely lost his interest in embryology, but his attention shifted in the early 1900s to the new science of genetics. Morgan's lab was no ordinary place. Nicknamed the 'Fly Room', it became home to thousands of generations of the common fruit fly (*Drosophila melanogaster*). The fruit fly is a convenient experimental animal. These flies have only four chromosomes in the nuclei of their cells, and it was the role of the chromosomes that Morgan wanted to understand: how important were chromosomes in passing on hereditary traits? The fruit fly chromosomes are large and easy to see on microscopic slides. Fruit flies also breed very quickly – leave out a plate of fruit and watch what happens. Many generations can be studied in a short space of time, to see what happens when flies with certain characteristics are bred with other flies. Imagine doing this kind of work with elephants and you can see why they chose fruit flies.

Morgan's fly room became famous, attracting both students and other scientists. It was a forerunner of the way much science is done today: a group of researchers working under a 'boss' – Morgan – who helps define the problems. The boss supervises the work of his or her team of younger researchers, who do the actual experiments. Morgan encouraged everyone to talk and work together so it has been hard to sort out exactly who did what. (When Morgan won his Nobel Prize, he shared the money with two of his younger colleagues.)

Almost by chance, Morgan made a crucial discovery. He noticed that one fly from a recent hatching had red eyes, rather than the usual white ones. He isolated this fly before breeding it with ordinary white-eyed flies. When he looked at the red-eyed offspring of that fly, he discovered first that all his red-eyed flies were female. That suggested that the gene was carried on the sex chromosome, the chromosome that determines whether the offspring will be male or female. Second, the inheritance patterns of eye colour followed the same rules as Mendel's peas – the eyes were either white or red, but never pink, or some colour in between. Morgan looked at other patterns of the tiny flies' inherited traits, such as wing size and shape. He and his colleagues examined their chromosomes under the microscope and began to develop maps of each chromosome, showing where the units of heredity (the 'genes', as they had been called) were located. Mutations (changes), such as the sudden appearance of the red eyes, could help locate where the gene was, as they carefully analysed what the chromosomes did during cell division. One of Morgan's students, H.J. Muller (1890–1967), discovered that X-rays caused faster mutations. Muller won his own Nobel Prize in 1948, and his work alerted the world to the dangers of radiation from atomic bombs and even from the X-rays being used medically. Morgan also showed that chromosomes sometimes exchange material when they are dividing. This is called 'crossing over', and it is another way in which nature increases the amount of variation in plants and animals.

Morgan and his group, as well as many others around the world, made genetics one of the most exciting sciences between about

1910 and 1940. The 'gene' was increasingly recognised as some material substance. Located on the chromosomes of the cells, the genes are passed, via a female egg fertilised by a male sperm, to the offspring, each parent contributing equally. Mutations were shown to be the thing that drove evolutionary change. They created the variation and they occurred naturally as well as by the artificial methods that Muller studied. The new genetics was central to evolutionary thinking. Even though what exactly the 'gene' was remained undefined, its reality was now beyond doubt.

This new genetic thinking had a darker side in society. If there was no soft heredity – so that eating better food, playing sports or being good, could not change the genes of your children – different methods would have to be used if you wanted future generations to improve. Darwin's 'artificial selection' had been practised for centuries, by livestock and plant breeders who tried to improve on the desirable characteristics of whatever they were breeding. Cows could be bred to yield more milk, tomatoes to be even juicier. In 1904, Francis Galton (Darwin's cousin) founded a 'eugenics' laboratory. He had coined the term 'eugenics', meaning 'good birth'. Here he had tried to change the reproductive habits of human beings. If intelligence, creativity, criminality, insanity or laziness could be shown to run in families (and Galton believed they could), it made sense to encourage the 'good' to have more children ('positive' eugenics), and to prevent the 'bad' from having so many ('negative' eugenics). Positive eugenics was the most common form in Britain. Campaigns encouraged educated middle-class couples to have more children, on the assumption that these couples were somehow 'better' than a casual labourer and his wife. In the late 1890s, the government had been frightened by the poor condition of recruits for the Boer War in South Africa. A large number of volunteers were rejected as physically unfit, unable even to carry a rifle. Then the First World War, from 1914 to 1918, saw mass slaughter in the battlefields of Europe. Many assumed that it was mostly the best who had been lost. Every nation throughout the Western world worried about the quality and strength of its population.

Negative eugenics was more sinister. Many assumed it was sensible to lock away people who were mentally disturbed or 'subnormal', criminals, even disabled people and others at the margins of society. In the USA, many states passed laws enforcing sterilisation, to prevent these people from having children. From the 1930s until their defeat in the Second World War in 1945, the Nazis in Germany practised the worst atrocities. In the name of the State, they first incarcerated, and then murdered millions of people they decided were unfit to live. Jews, Gypsies, homosexuals, the mentally disturbed or deficient, criminals: all were herded up and either sent to concentration camps or executed.

The Nazi period made 'eugenics' a dirty word. As we shall later see, some people believe that eugenics could return through the back door, as scientists learn more and more about what we inherit, and how it affects who we are. We all need science, but we must all make sure that it is used for good.

CHAPTER 35

Where Did We Come From?

Today we know that we share 98 per cent of our genome with our closest animal relatives, the chimpanzees. That's an awful lot of similarity, but there are some crucial differences. While chimps do communicate they don't talk together as humans do. And we can read and write. Take a step back and we find that humans and chimpanzees, together with the gorillas and orang-utans, make up the family of *Hominidae*, often known as the 'great apes'. We humans are less closely related to gorillas and orang-utans, but at some point in the past all four of these groups shared a common ancestor, from which each group evolved. That was a long time ago, perhaps fifteen million years.

We find our great-ape 'cousins' fascinating and slightly disturbing. Those who wrote about them and studied them in the past did so too. They wondered where this brute animal, that seemed so like us and yet so different, fitted into creation. In 1699 an English anatomist, Edward Tyson (1651–1708), obtained the body of a dead chimpanzee. He carefully dissected this exotic animal and compared what he found with what he knew of human anatomy. It was the first time anyone had looked so closely at a chimpanzee. Tyson slotted it into Aristotle's Great Chain of Being – just below us. It was natural, he argued, that some animal would smooth over the gap between humans and the rest of the animal kingdom. He didn't say it, but Tyson had suggested the need for a 'missing link' in the chain, something that connects us to other animals.

In Britain, Germany and France, a growing number of human artefacts such as flint arrows and axe heads were being uncovered. This was exciting evidence of human presence going back millennia. These tools were often found in caves and fossil sites among the fossilised remains of extinct animals – the fearsome sabre-tooth tigers and giant woolly mammoths. These extinct animals and the Stone Age humans who had made the tools had obviously been alive at the same time. Humans had been on the earth for tens of thousands of years ... not the much shorter period that most people believed. Not everyone agreed, of course, but Darwin's friend Thomas Henry Huxley (1825–95) had no doubts. Huxley was excited by the discovery in 1856 of 'Neanderthal man' in a cave in the Neander Valley in Germany. He wrote about this fossil, and about modern humans and the great apes in his book *Man's Place in Nature* (1863). We know now that this was the first fossil hominin that did not belong to our species, *Homo sapiens*, the biological name that Linnaeus gave us (Chapter 19). *Hominin* is the name now used for ourselves and for our extinct ancestors, and as more fossil evidence is uncovered, the group gets larger. The tree of life is growing, and gradually being filled in.

At the time, Huxley was cautious enough to recognise that a single find doesn't tell you everything about a whole species, and so he kept Neanderthal man in the same species as modern humans. But he was confident that this was a very old specimen, one that had been around long enough for evolution to have taken place. There had certainly been some changes, for although Neanderthal man was similar enough to us, he was also different.

The skull had immense brow ridges and a much larger cavity for the nose. The proportions of the limbs and body were different to ours. It was even possible that this was a deformed body rather than another species. In time we would learn that the Neanderthals were the first hominins to bury their dead.

Huxley knew all about Darwin's ideas on human evolution before the great man published two books in quick succession laying out his ideas and evidence for our ancestry. In 1871 *The Descent of Man* did what Darwin had avoided doing in *On the Origin of Species*: it focused his compelling account of our world upon the human race. In 1872, his book *The Expression of the Emotions in Man and Animals*, added an important psychological dimension to his argument. He based the book on his careful watching of his own children, their smiles and grimaces, among many other behaviours. Humans were part of life on earth, like all the other species of plants and animals. Darwin concluded that our ancestors had probably lived in Africa, where humans had first evolved.

Darwin's depiction of evolution as a 'tree of life' meant that we could not be descended from modern apes. But it was the 'ape man' connection that immediately caught the public's imagination. His ideas on evolution were first debated in public at a crowded meeting in Oxford, organised by the British Association for the Advancement of Science. The Association aimed to bring the latest scientific knowledge to everyone and held a meeting every year where scientists talked and debated what was new. The meeting in 1860 was full of drama, so sensational was the 'ape man' idea. The discussion of Darwin's ideas on evolution was eagerly awaited, with Bishop Samuel Wilberforce leading the anti-Darwinians and Huxley the pro-Darwinians. Wilberforce, thinking he was clever, asked Huxley whether he was descended from the apes on his grandfather's or his grandmother's side. Huxley replied that he would indeed rather be descended from an ape than waste his time and brain on such a silly question: Wilberforce had quite missed the point. Wilberforce remained unconvinced, but Huxley and evolution came out on top that day.

The discoveries of mankind's long existence on the earth encouraged naturalists, anthropologists (who study humankind) and archaeologists to ask the question: What had been the original condition of human beings? 'Cave men' emerged in this period from the discoveries in caves in Britain and Europe. It was clear that these cave dwellers had used fire. Weapons, stone tools and cooking utensils were all found. Anthropologists and explorers also discovered hunter-gatherer groups in Africa, Asia and South America, and suggested that all human societies had passed through common stages of social development. E.B. Tylor (1832–1917) became the first professor of anthropology at Oxford. He used an idea of 'survivals' to put forward a grand path of human social and cultural evolution. By this he meant social and religious practices, superstitions and different ways of organising family relationships. According to Tylor, these survivals were frozen in the 'primitive' people of Africa, for instance, and gave clues to the common past of humankind. Tylor and others wanted to understand the origins of language and looked at gestures and other ways of communicating.

This early anthropology contrasted a dynamic Europe, North America, Australia and New Zealand with the presumed unchanging lives of 'primitive' peoples, or even the long-established and complex cultures of India and China. We now see it as arrogant. Applied to Western society, the idea of evolutionary competition and struggle seemed to explain why some individuals prospered and some didn't. As industrial capitalism gained strength, 'social Darwinism' – evolution applied to human culture – began to be used to explain why some people were rich and others were poor, and some nations powerful and others not. Social Darwinism justified the triumph of strong individuals, races or nations over weaker ones.

While some people were debating social Darwinism, others were discussing biological evolution. Until the 1890s, all the fossilised human remains that were discovered were considered to be *Homo sapiens*. The status of Neanderthal man remained uncertain. Then, a Dutch anthropologist, Eugène Dubois (1858–1940) went to the Dutch East Indies, looking for evidence of human evolution in the land of the orang-utan. In Java (now Indonesia), he found the top of a fossilised skull belonging to a non-human creature that had walked upright. He called the creature 'Java man'. Attention turned to Asia, as the place where humans must have evolved. Java man, along with another old human skeleton found in France at Cro-Magnon, stimulated questions about what had happened first. Was it walking upright, on two legs? Or a large brain? Or language and living in societies?

There have been many more discoveries of pre-human hominins in Asia. But in the twentieth century, it was Africa that proved how shrewd Darwin's prediction had been. In 1924, a fossil was discovered by the Australian anatomist Raymond Dart (1893–1988). It became known as the 'Taung child', and its significance was championed by the South African doctor Robert Broom (1866–1951). Tuang child had teeth like a human child but its brain was too ape-like to be considered human. Broom believed that Dart's fossil (and several more found subsequently, including an adult) was an ancient ancestor of human beings. Dart named it *Australopithecus africanus*, literally, the 'southern ape of Africa'. We now think it is between 2.4 million and 3 million years old. After Taung child, Africa yielded many other important fossils, helping piece together man's evolutionary ancestry. Louis and Mary Leakey (1903–72; 1913–96) made the human story even more famous. They were working in the 1950s mainly in Olduvai Gorge in Kenya, and Louis Leakey stressed that early hominins were tool-makers. He called one of the fossil hominins that had lived 1.6 to 2.4 million years ago *Homo habilis* – the 'handyman'. Mary Leakey discovered in the 1970s some footprints that were 3.6 million years old, preserved in volcanic ash that had hardened. The footprints were of three upright hominins, along with other animals, and suggested that walking on two feet came first, before hominins evolved with a big brain.

For the first half of the twentieth century, the study of human fossil bones was complicated by some curious finds in a gravel pit in the village of Piltdown, East Sussex, in southern England. The discoveries began in 1908. Then in 1912 a local amateur archaeologist, Charles Dawson (1864–1916), announced the recovery of a skull at Piltdown. The find generated tremendous excitement. 'Piltdown man' had a modern-looking human skull with a jawbone that was ape-like. It looked like a real missing link, a kind of 'ape man'. A number of eminent scientists published papers on the strange fossil. But it was difficult to fit it into the emerging sequence of the new hominin and ancient ape fossils. Piltdown had always seemed fishy, and in the early 1950s dating techniques that had not been available in 1908 proved that it had been a huge forgery. Piltdown man combined a modern human skull with the jaw of an orang-utan, soaked in chemicals to make them look old. The teeth had also been filed down. No one is sure 'whodunnit' – there are several suspects but no definitive conviction. Dawson himself is high on the list of suspects.

With Piltdown revealed as a hoax, the other fossil hominins could be placed in a more likely order, using radiometric dating to learn their age, and comparing their physical characteristics. One fossil in particular, nicknamed Lucy, has become a celebrity, going on tour and having her 'biography' written. Lucy was uncovered in Ethiopia in 1978, and her skeleton was more than half complete. She had lived some three to four million years ago, long before the Taung child. Like the Tuang child, she is of the genus *Australopithecus* but is an earlier species, *afarensis* – 'ape of Afar'. Lucy's legs, pelvis and feet mean she could probably have walked upright and climbed in trees or on rocks. Her brain cavity was not much bigger than a modern chimpanzee's, but her brain was larger than a chimp's, in relation to the size of her body. (The brain-to-body ratio is a better guide to mental functions than mere size: elephants have larger brains than humans, but smaller brain–body ratios. There are of course many other factors to 'intelligence' than simply brain size.) Lucy really did show 'mixed' characteristics, not yet even crudely 'human', but a successful creature in her own right.

Hundreds of fossil hominins from many parts of the world have allowed us to get a pretty clear idea of the evolutionary path that led to modern human beings. We can even tell what was eaten and what parasites infected our ancestors. The puzzle has many missing pieces, and there is much debate on details: what does this tooth tell us, or the shape of that thigh bone? There will be more surprises in store, too, because fossils are continually unearthed. In Indonesia in 2003, the Australian archaeologist Mike Morwood and his colleagues found fossils of small hominins on the island of Flores. They had lived as recently as 15,000 years ago, but are probably of an unknown species. The exact status of *Homo floresiensis* ('Flores man', nicknamed 'the Hobbit') is still uncertain. Attempts at DNA analysis (the most reliable way of establishing biological relationships) have so far been unsuccessful.

Working out how Neanderthals relate to modern humans is an exciting challenge, too. The species certainly lived at the same time as *Homo sapiens* in Europe, 50,000 or so years ago. We carry some of their genes. Did the coming of *Homo sapiens*, 'modern' man, contribute to the extinction of Neanderthals? We are not sure. Did they breed with each other? Probably. Both Neanderthals and *Homo sapiens* suffered from the very cold European temperatures the last time glaciers covered Europe, and the Neanderthals did not survive.

To reconstruct the human family tree from fossils of differing ages, and in different locations, we use the same tools and techniques as we do for other animals such as the horse or hippopotamus. Of course, there is much more emotion involved, when it's humans rather than hippos. But the evidence is there, and palaeontologists, anthropologists, archaeologists and other specialists continue to put the pieces together. They have used the evidence to work out that hominins, including, at last, *Homo sapiens*, first lived in Africa and spread from there. There is still much we don't know about the migrations of early hominins. Were there several movements out of Africa? What led to the rapid evolution of the large brain that sets our own species apart from our cousins? Science deals with the how, not the why. This seems especially true when we think about our ancestry and, as Huxley put it, 'man's place in nature'.

CHAPTER 36

Wonder Drugs

There may be five million trillion trillion bacteria on earth. That's 5×10^{30} or 5 with thirty noughts after it – an astounding number. Bacteria can live almost anywhere on earth: in the soil, the oceans, deep underground on rocks, in Arctic ice, in the boiling water of geysers, on our skin and inside our bodies. Bacteria do all sorts of useful things – without them what would happen to all the rubbish they digest? We benefit from that digesting trick too. The bacteria that live in our guts help us break down the food we eat to release the proteins and vitamins. Some bacteria even turned out to make useful drugs, along with some other micro-organisms, the fungi. Most of us have been prescribed some of these antibiotics.

In the nineteenth century, scientists had discovered how harmful some bacteria are, causing disease and infecting wounds. Chapter 27 tells the story of how their 'germ theory' of disease became accepted. Straight away, they began looking for drugs that could kill the invading bacteria without harming the cells of the body. It was a quest for 'magic bullets', said the German doctor Paul Ehrlich (1854–1915) He came up with a drug to treat syphilis, but it contained arsenic, which is poisonous, so it had to be used very carefully and had serious side effects.

In the mid-1930s, the German pharmacologist Gerhard Domagk (1895–1964) began to use the chemical element sulphur. (Pharmacology is the study of drugs.) He produced a compound called Prontosil, which was effective against several kinds of disease-causing bacteria. One of the first experimental patients was his daughter, whose hand had become infected with *Streptococcus*, a nasty bacterium that causes infections of the skin. Doctors had said that the only way to try and save her from the life-threatening infection was to amputate her arm. Prontosil successfully cleared up the infection. It was also effective against scarlet fever and a fatal bacterial infection called puerperal fever, which killed women after they had given birth. Prontosil began to be widely used from 1936 and contributed to a dramatic fall in the number of these deaths. It and other sulphur-containing drugs were among the best drugs doctors could prescribe against certain bacteria. Domagk won a Nobel Prize in 1939 (though at the time the Nazis forbade Germans from accepting it).

The next Nobel Prize for the discovery of a drug came in 1945. Three men, the Scot Alexander Fleming (1881–1955), the Australian Howard Florey (1898–1968) and the German refugee Ernst Chain (1906–79), shared the prize for the discovery of penicillin, the first 'antibiotic' drug. An antibiotic is a substance produced by one micro-organism that can kill other micro-organisms. It harnesses for our benefit something that happens in the natural world all the time. Penicillin was purified from a natural source, the micro-organism *Penicillium notatum*, a mould or kind of fungus. You can see small rings of blue fungi growing on old, mouldy bread. If you like to eat mushrooms, you are of course eating another kind of fungus. There are thought to be 1.5 million species of fungi on our planet. They have complex life-cycles including a spore stage, which is similar to the seeds of plants. Today antibiotics can also be created in the laboratory rather than from a natural source, but it's the same basic idea.

Penicillin's story begins in the 1920s. Like all the best stories, there are several versions. One has it that in 1928 a spore of the mould drifted through an open window in Alexander Fleming's laboratory at St Mary's Hospital in London. What he noticed was that some of the bacteria he was growing on a Petri dish stopped growing where the spore had landed. He identified the spore as coming from *Penicillium*, did more work with it and published his results to share them with other bacteriologists. But he couldn't see how to make enough of whatever the spore had produced to be of any use. So he left it as a curious, possibly promising, laboratory observation.

A decade later, Europe was plunged into the Second World War. War always brings outbreaks of infectious diseases, among soldiers and civilians alike. So the pathologist Howard Florey, who had settled in England, was asked to look for effective drugs against infections. One of his associates, Ernst Chain, began reading everything he could find, including Fleming's old paper. Next he tried extracting the active substance produced by the penicillin mould. In March 1940, their laboratory assistant, Norman Heatley (1911–2004), found a better way of obtaining this promising substance. Working in difficult wartime conditions, they had to make do with few resources, using bedpans and milk churns as containers for growing the solutions of mould. Nevertheless, they obtained some relatively pure penicillin. Tests on mice showed that it was very effective in controlling infections. Purifying the miraculous substance was extremely difficult: it took a tonne of a crude solution of penicillin to produce two grams of the drug. Their first patient was a policeman who had become infected after a scratch from a rose thorn. When given the drug, his condition improved briefly. They filtered his urine to recover the precious drug, but he died when the supply ran out.

Wartime Britain did not have the industrial resources to produce enough penicillin. So in July 1941, Florey and Heatley flew to the USA to encourage American pharmaceutical companies to take this on. Florey was an old-fashioned scientist. He believed that discoveries such as theirs were for everyone's good and should not be patented. (Patents are a way of protecting inventors' ideas and making sure that no one else can copy them.) The Americans had other ideas. Two companies in particular developed special methods of producing penicillin on a vast scale. To make back all the money they had invested in the research, they took out patents, which meant that no one else could use their methods to make the drug. By 1943, penicillin was available for military and some civilian use. It was shown to be effective against the *Streptococcus* bacterium, as well as some of the organisms that cause pneumonia, a lot of wound infections and some sexually transmitted infections. Soon, enough was being made to ensure that those who could be treated would live, when otherwise many would have died, especially the soldiers fighting to end the war.

While Florey and his team were busy with penicillin, Selman Waksman (1888–1973) was working on the antibiotic properties of bacteria. Waksman had come from Ukraine to the United States in 1910. He was fascinated by the micro-organisms that live in the soil, and had seen how some of these micro-organisms killed other bacteria in the soil. From the late 1930s, he tried to isolate compounds from these bacteria that could act as antibiotics. With his students, he isolated some effective substances, but they were too toxic to be used in humans. Then, in 1943, one of his students isolated *Streptomyces*, and the drug streptomycin was made from it. It proved effective and not too harmful to patients. Amazingly, it worked against the bacterium that causes tuberculosis, that deadly disease that had killed more people than any other disease during much of the nineteenth century. Although it was less common in the West by the 1940s, it was still taking its toll everywhere. Its victims were often young adults, leaving loved ones bereft, and children without their parents.

Penicillin and streptomycin were just the beginning of a whole range of antibiotics and other chemicals that cured infectious diseases. In the years after the Second World War, they made people very optimistic about the power of medicine to combat and even eradicate such disease. Fewer people in the West died from infections, and with the exception of new infections such as AIDS, this has continued. Without doubt, many young people in the twenty-first century can live healthier lives than their parents or grandparents.

But if the optimists of the 1960s had looked carefully at the story of an earlier 'miracle drug', they might have realised miracles are unlikely. That earlier drug was insulin, used to treat diabetes since the 1920s. Diabetes is a horrible affliction. If it is not treated, the body wastes away, its victims become painfully thin, are always thirsty, urinate frequently and eventually sink into a coma before dying. It mostly affected young people, who died within a couple of years. It is a complicated disease, but the special cells that produce insulin naturally in the pancreas – an organ near the stomach – stop doing their job. Insulin is a hormone, a chemical 'messenger', and it keeps the correct amount of sugar (glucose) in our blood.

While penicillin originated in a lucky chance, the story of insulin is one of painstaking research into how some parts of the body work. Researchers had already shown the role of the pancreas by removing it from dogs (or other animals) that then suffered a diabetes-like illness. Over the summer of 1921, at the University of Toronto, Canada, Professor J.J.R. Macleod (1876–1935) was away. A young surgeon called Frederick Banting (1891–1941) and his medical student assistant Charles Best (1899–1978) conducted a series of simple experiments. With the help of a biochemist, James Collip (1892–1965), they managed to extract and purify insulin from the pancreases of dogs. When they gave the insulin to their experimental animals that had had their pancreases removed, they recovered from their diabetes.

Insulin was described as a 'force of magical activity'. It could literally bring the victims of this kind of diabetes back from certain death. One of them was fourteen-year-old Leonard Thompson, the first person treated with insulin injections in 1922. Leonard was severely underweight and was confined to a hospital bed because he was so weak. The injections brought down his blood sugar towards normal levels, he gained weight, and was able to leave hospital with his syringe and insulin supply.

One year later, Banting and Professor Macleod were awarded the Nobel Prize, and shared the prize money with Best and Collip. Such speedy recognition showed how important everyone considered their work to be. Insulin *was* very important. It offered years of extra life to many young people who would otherwise have died. What it didn't offer was a *normal* life. Diabetics had to monitor their food, give themselves regular insulin injections, and frequently test their urine for sugar. This was much better than nothing. But a decade or two later, many of these early diabetics began to suffer from other health problems: kidney failure, heart disease, difficulties with their eyesight and painful ulcers on their legs that refused to heal. Insulin changed an acute fatal disease into a lifelong problem to be managed forever. The same problems also apply to the other kind of diabetes, which occurs mostly in overweight adults and is called Type II diabetes. It is now the most common form, and more and more people suffer from it. Modern diets contain too much sugar and refined foods, and obesity has become a global epidemic. Medical science has helped: pills can lower the blood sugar. But Type II diabetics face the same kind of problems in later life. Medicine is simply not as good as our own natural systems at regulating the level of sugar in our bodies.

Nature has shown us that we can't rely on penicillin and other antibiotics. These drugs are still useful, but the bacteria that cause disease have adapted to them. Darwin's discovery of natural selection applies throughout nature, and many bacteria have developed defences against the antibiotics that used to kill them. The *Staphylococci* and the tubercle bacillus (which causes tuberculosis) have shown themselves to be especially adaptable. Like all other living creatures, their own genes sometimes mutate, and the mutations that help them to survive are the ones that pass on to the next generation. Treating infections has now become a kind of cat-and-mouse game: developing new drugs to attack germs that evolve to resist almost anything we can throw at them. One recent problem is MRSA (methicillin-resistant *Staphylococcus aureus*). *S. aureas* is one of those bacteria that normally lives on our bodies, even if it may cause the usual slight infection after a scratch. Its antibiotic-resistant form is dangerous. It is commonly found in hospitals because so many antibiotics are used there, and the bacteria that do survive are often those that have developed resistance. And it is not just bacteria that fight back against our attempts to control disease. Some of the parasites that cause malaria are resistant to almost all the drugs we have.

We now know that bugs tend to develop their resistance when patients don't finish taking the full course of their medicine, or when the wrong dose is given. It also happens when the drugs are misused: antibiotics are often given to patients inappropriately, for infections, colds or sore throats caused by viruses. (Antibiotics fight *bacteria*, and can do nothing against viruses.) If your dose of antibiotics is not enough to kill the disease-causing bacteria, the treatment can instead help resistant bacteria to survive. Those bacteria might in the future cause untreatable disease.

Despite all these problems, doctors have many more powerful and effective drugs than ever before. Some, like insulin, control rather than cure the disease, but all these modern medicines have given people in the 'developed' world the chance to live longer lives. In many countries in the 'developing' world, too, life expectancy has also risen. But there, serious problems remain: it is not always easy to see a doctor, get enough to eat, drink clean water, or live in a comfortable home. Since the early 1990s, the gap between rich and poor has widened in rich countries, and has widened too between the rich and the poor countries. This shouldn't be.

Today it costs a lot of money to provide medical care. We use a lot of clever technology to diagnose illness and then treat it. Developing and testing new drugs now takes much more money than it did to produce penicillin. So we need to look after ourselves if we can. No matter how amazing the medicines, it is still true that 'prevention is better than cure'.

CHAPTER 37

Building Blocks

As time went on, scientists tended to specialise in their chosen fields. Still, biologists traditionally did biology, chemists did chemistry and physicists did physics. So what was happening in the 1930s, when first chemists, and then physicists, decided it was time for them to take on the problems of biology? Chemistry was about how substances combine and react. But it was becoming clear that living organisms – the biologists' subject – were made up of some of the elements of the chemists' periodic table, such as carbon, hydrogen, oxygen and nitrogen. Physics was about matter and energy, which by this time was full of atoms and their sub-atomic particles. Wasn't that a way of understanding more about the chemists' elements? To sum up, couldn't chemistry and physics explain living organisms as a series of chemical reactions and atomic structures? And might that provide an answer to one of the oldest questions in science: What is life?

In the early decades of the twentieth century, Thomas Hunt Morgan had used his little fruit flies to show that it was the chromosomes in the cell's nucleus that carried the stuff of heredity. 'Stuff' was a good word for it. Geneticists had got very good at showing what this stuff did. They could show how, on different bits of a chromosome, the different genes could result in the development of an eye or a wing. They could even show how mutations produced by X-rays could lead to unusual wing shapes because, they believed, they affected the genes. But they didn't know what a gene was.

Could proteins be this genetic stuff? Proteins are fundamental to many of the reactions that go on inside our bodies. The proteins were the first group of compounds to be systematically studied by molecular biologists. As the name suggests, molecular biology is a science that seeks to understand the chemistry of the molecules in living things, and how they work. Proteins are mostly very large, complex molecules. They are composed of groups of amino acids, which are smaller and simpler compounds than proteins. Being simpler, it was easier to find out what the amino acids were made of, using ordinary chemical analysis and synthesis. About twenty amino acids are the building-blocks that in different combinations make up all of the proteins in plants and animals.

How these amino acids fit together to make the proteins was a much more difficult question. This was where physics began to play a part – it turned out that X-rays provided clues. The first thing was to make a crystal of the protein you wanted to study. Next, you bombarded the crystal with X-rays. As the X-rays hit the crystal they would be bent as they passed through it, or would be reflected back in a particular pattern, known as the diffraction pattern. It could be caught on a photographic plate.

Reading the patterns captured on the photographic plate is a tricky business. What you see is an intricate picture of lots and lots of dots and shadows. You are looking at a flat, two-dimensional image but you have to think in three dimensions, and just putting on 3D glasses won't help. As well as being able to visualise the picture, you also need to know your chemistry and understand how elements join together. And be good at maths too. Someone who took on this challenge was the chemist Dorothy Hodgkin (1910–94) who worked at Oxford University. We partly owe what we know about the structure of penicillin, Vitamin B12 and insulin to her research in X-ray crystallography. She won her Nobel Prize in 1964.

Linus Pauling (1901–94) was also good at using X-rays to work out the structure of complex chemical compounds. In a brilliant series of experiments, he and his colleagues were able to show that if just one amino acid was missing from the haemoglobin molecule in our red blood cells, it produced a serious disease: sickle-cell anaemia. (Rather than being round, the red blood cells that contain this haemoglobin are shaped like a sickle.) This molecular flaw is found mostly in Africa, where malaria is always present. It is now understood to benefit the people who have the flaw, because the sickle cells help to protect against the most serious form of malaria. This is an example of human evolution in action. People with only the trait (a single gene, inherited in the way Mendel first studied in peas) are moderately anaemic, but they are more resistant to malaria. Individuals who inherit the sickle-cell gene from both parents are seriously ill from anaemia. The symptoms of sickle-cell anaemia had been identified early in the twentieth century. Fifty years later, Pauling used the new techniques of molecular biology to understand what was going on, and his research began a new era in medicine: molecular medicine.

After his success with proteins, Pauling almost achieved the biggest prize: revealing the molecular structure of the genes. His X-ray experiments showed that many proteins, such as those that make your hair and muscles, or carry your oxygen on the haemo-globin molecules, have a special shape. They were often wound into a spiral (helix). By the early 1950s, lots of scientists thought that the genes were made up of deoxyribonucleic acid. This compound is much better known as DNA, and a lot easier to say. DNA had been discovered in 1869 but it took a long time to understand what it might do and what it looked like. In 1952 Pauling suggested that it was a long coiled molecule made up of three strands twisted together – what was called a triple helix.

While Pauling was at work in California, two groups in England were hard on his heels. At King's College, London, the physicist Maurice Wilkins (1916–2004) and the chemist Rosalind Franklin (1920–58) were turning themselves into molecular biologists. Franklin was particularly good at producing and reading the photographs produced by X-ray crystallography. At Cambridge, a young American, James Watson (b. 1928), had given up his earlier interest in ornithology (the study of birds) and teamed up with Francis Crick (1916–2004). Crick had studied physics, and after working as a physicist for the Admiralty during the Second World War, he had gone back to university as a mature student, this time to study biology. Watson and Crick would become one of the most famous double acts in science.

Crick shared his experience on the X-ray analysis of the structure of proteins. He and Watson knew that DNA is found on the chromosomes in the cell nucleus – the same cell components that Morgan had analysed thirty years before. They made paper cut-outs and built models to help them see possible structures of DNA. They also benefited from the photographs that Franklin had produced. Early in 1953 they created a new model that matched all the X-ray data. This one, they said, was the right one. Celebrating in the pub that night, the story goes that they claimed they had discovered 'the secret of life'.

If the other drinkers that night were a bit in the dark, wondering what they meant, readers of the weekly scientific magazine *Nature* would soon find out. Crick and Watson published their findings in the issue of 25 April 1953, which also included a paper by the London team of Wilkins and Franklin. But it was Crick and Watson who showed that DNA is made of two twisted strands, not three as Pauling had said. The strands were joined together by cross-pieces – so that it looked like a long flexible ladder twisted into a spiral. The uprights on the ladder are a kind of sugar – the D or *deoxyribo* part of the molecule and phosphates. Each rung of the ladder is made of a pair of molecules: either adenine with thymine, or cytosine with guanine. These became known as the 'base pairs' of molecules. So if that was the structure, how did it explain 'the secret of life'?

The base pairs are joined together by hydrogen bonds. When cells divide, the coils unwind, almost as if they are 'unzipping'. The two halves now present the templates for two identical chains to be made by the cell. So Watson and Crick had shown how genes could be passed from parent to offspring and how 'daughter' cells would contain the same set of genes as the original 'mother' cell. It was simple and elegant, and it immediately seemed obvious. In 1962, when the scientific community had fully accepted the structure and role of DNA, Crick, Watson and Wilkins shared the Nobel Prize. Only three people can officially share a Nobel Prize. But Rosalind Franklin was not ignored: she had died of ovarian cancer, aged only thirty-eight, in 1958.

Francis Crick went on, with others, to explain why genes are so important for living organisms, besides their role in inheritance. What genes do in their daily activity is make proteins. The 'genetic code' is made up of three neighbouring rungs on the ladder, and each triplet of rungs (the 'codon') is responsible for a single amino acid. Crick showed how little portions of the DNA molecule provide the codes for the amino acids that make up proteins such as haemoglobin and insulin. Geneticists realised that the order of the base pairs within the DNA molecule is crucial, because that determines which amino acids will be built into the proteins. Proteins are very complex molecules, sometimes with dozens of amino acids, so a long sequence of DNA is necessary to make such a protein.

With the basic workings of DNA understood, scientists could now make sense of the kinds of thing that Morgan had seen in his fly room. Morgan had been looking at the visible characteristics of whole organisms – in his case, the fly with its normal white eye or mutant red eye. This kind of visible trait is called a *phenotype*. From now on, scientists could begin to work at a level below the whole organism, at the level of the genes – what now became known as the *genotype*.

Discovering the structure of DNA was a huge turning point in the history of modern biology. It showed that biologists could understand things in terms of the molecules in the cells, previously the domain of chemists. This was now what everyone wanted to do. Later research revealed that the amino acids and then the proteins were made in the cell's cytoplasm – the liquid bit outside the nucleus. Learning how this little protein factory worked included the discovery of RNA. This is ribonucleic acid, similar to DNA, but with only one strand, not two, and a different kind of sugar. The RNA had an important part to play in the flow of information from the DNA in the nucleus of the cell to the protein factory in the cytoplasm.

Molecular biologists were to transform our knowledge of how diseases originate. They uncovered how proteins like the hormone insulin did their job in regulating blood sugar. They gained a better understanding of cancer, one of our most feared modern diseases. Although all cancers can overwhelm the whole body, and thus become a general disease, they start with a single mutated cell, which misbehaves and doesn't stop dividing when it ought to. These runaway cells are greedy. They use up the body's nutrients, and if they spread to a vital organ, the cancer cells disrupt its functions, leading to further illness. Finding out how this happens at the molecular level was essential before better drugs could be developed to slow the process down, or even stop it.

Studying these dynamic processes is difficult in large, complicated animals like humans, so much of the work of molecular biologists depends on using simpler organisms. A lot of the early research on the actual functions of DNA and RNA was done with bacteria, and cancer research uses animals such as mice. Translating these findings to human beings isn't easy, but that is the way modern science operates: going from the simpler to the more complex. This method has helped us understand the processes that have driven evolution for millions of years. It turns out that DNA is the molecule that controls our destinies.

CHAPTER 38

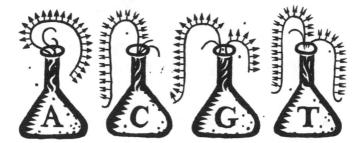

Reading 'the Book of Life'
THE HUMAN GENOME PROJECT

Humans have about 22,000 genes (the exact number is history in the making). How do we know this? Because scientists in laboratories all over the world collaborated on the Human Genome Project. This hugely ambitious project counted our genes by using DNA sequencing, and answered a question left hanging when Crick and Watson revealed the structure of DNA. The 'sequencing' meant the position, on the chromosomes, of every one of the three billion 'base pairs' of molecules that make up our genome. That's an awful lot of molecules of adenine and thymine, cytosine and guanine arranged in their double helix in the nucleus of each of our cells.

If understanding DNA had given us 'the secret of life', the Human Genome Project was about reading 'the book of life'. For that is what your genome is, the genes for everything about you, from the colour of your hair to the shape of your little toe. It is also about things that cannot so easily be seen: the instructions for one fertilised egg cell to become two and then four and all the way up to a whole baby in the womb. It controls the biological programmes in cells that produce proteins like the hormone insulin to regulate our blood sugar. It runs the programmes for chemicals in the brain that transmit messages from one nerve to the next.

The Human Genome Project began in 1990 and was supposed to be finished by 2005. But in a moment of science drama, on 26 June 2000, five years ahead of schedule, an unusual thing happened. Amid great fanfare, on live television, the President of the United States of America and the Prime Minister of Great Britain announced that the first draft of the project had been completed. They were accompanied by some of the scientists who had done the work, but the presence of these two world leaders was an indication of just how important it was to understand the genome.

It would take another three years, until 2003, to produce a much better version of this book of life – filling in the big gaps and correcting most of the errors. Even so, that was two years sooner than originally planned. During the years of the project the methods and technology used by the scientists, particularly the assistance provided by computers, had also advanced.

The genome project had developed from decades of research that followed the discovery of DNA. After Crick and Watson's revelation in 1953, an important thing to do was to 'clone' strands of DNA, to get more of the particular part of the DNA molecule you wanted to investigate. In the 1960s molecular biologists worked out that this could be done using enzymes and bacteria. Enzymes are proteins that can do all sorts of things depending on their individual structure. They were used here to do one of their natural jobs: cutting DNA into little sections. These little sections were then inserted into bacteria in a special way. Bacteria reproduce very quickly, and as these modified bacteria reproduced they also made copies of the added sections of DNA. These copies, the clones, could then be harvested for further research. The process created a lot of excitement but it was only a beginning. Whole cells as well as bits of DNA can be cloned. A sheep called Dolly was the first mammal to be cloned from an adult sheep cell. She was born in 1996 and died in 2003. Cloning techniques continue to develop and are one of the most newsworthy areas of molecular biology research.

Now that the scientists had lots of the bits of DNA to experiment with, they began to try to solve the problem of DNA sequencing: to reveal the order of the base pairs of molecules in DNA. This was a job for the English molecular biologist Frederick Sanger (b. 1918), working in Cambridge. Sanger had already won one Nobel Prize in 1958 for working out the order of the amino acids of the protein insulin.

One of the key differences between amino acids and DNA is that the DNA molecules are much longer, and have many, many more base pairs than proteins have amino acids. Also each amino acid is less chemically similar, whereas the DNA bases were much like each other, which makes them harder to sort out. Building on his own earlier work, and that of others, Sanger found a way to prepare short strands of DNA using radioactive labels, chemicals and enzymes. He adapted various biochemical methods to find a way of separating out the adenine, thymine, cytosine and guanine from each other. To do this, he exploited the fact that as chemical compounds they have slightly different chemical and physical properties. The best results came with a process called *electrophoresis*.

To make sure the results were accurate enough, Sanger and his team processed multiple copies of each strand several times and compared the results. It was a very time-consuming, repetitive process. But by using lots of the short strands of the long molecule and then looking to see where they started and ended, they managed to match up the strands and produce a readable DNA sequence. In 1977 they had their first success in reading the genome of an organism. It was a humble one, a bacteriophage called phi X 174. Bacteriophages are viruses that infect bacteria, and phi X 174 was one often used as a tool in molecular biology laboratories. In 1980 Sanger won his second Nobel Prize for this valuable work.

The next genome targets were also laboratory organisms. Despite how hard it was to produce a readable DNA sequence, molecular biologists carried on with their research. Meanwhile, innovations in computing helped with analysing the patterns of the bases on the short strands. The scientists pressed on keenly. If they knew exactly which genes an organism had, and which proteins each gene could manufacture, they would be able to understand very basic things about how the organism was made, literally cell by cell from fertilised egg to adult.

The fruit fly was an obvious candidate for their research. Thomas Hunt Morgan and his group had already done a lot on its inheritance patterns, and some crude gene-mapping, before 1950. Another was a tiny roundworm called *Caenorhabditis elegans*. At only one millimetre long, it had exactly 959 cells, including a simple nervous system. Now it might not seem like much of a pet, but *C. elegans* was the favourite laboratory animal of Sydney Brenner (b. 1927), and had been for many years. Brenner had come from South Africa to the Laboratory of Molecular Biology (LMB) in Cambridge in 1956. Since the 1960s he had been investigating its development, since its cells were easy to see. He thought it would be possible to determine exactly what each of the cells in the embryo worm would become in the adult. He hoped that if he could reveal the worm's genome, he would be able to relate its genes to how the adult worm carries out its living functions.

In the course of their work, Brenner and his team also learned a lot about the ordinary lives of cells in an animal, including one very important job that the cell must do: die when it is time to die. Plants and animals always make new cells: think of your skin and how it rubs off when you have been in the bath a long time. We get rid of the dead stuff, and new, living cells replace it underneath. All this living and dying within an organism is a regular feature of nature, and the genes programme this process. That is why cancer cells are so dangerous: they don't know when it is time to die. Trying to influence the gene that has failed to tell the cell it is time to stop dividing is a major part of modern cancer research. Brenner and two colleagues won the Nobel Prize in 2002 for their work with the lowly roundworm.

By this time, one of those colleagues, John Sulston (b. 1942), was leading the British team taking part in the Human Genome Project.

The project stands as a symbol of modern science. First, it was expensive and thousands of people worked on it. The modern scientist is rarely a lone worker, and it is quite normal today for scientific papers to have dozens or even hundreds of authors. The work may require many individuals with different skills. It's been a long time since William Harvey worked alone on the heart, or Lavoisier in his laboratory had his wife as his only assistant. Several laboratories worked together on sequencing the human genome. They divided up the chromosomes between them, so cooperation and trust were needed, and every lab had to produce the sequences to the same high standards. This needed many smaller portions of the DNA, and then computer analysis to fit them together in a single sequence. Running these laboratories was expensive, so generous funding was needed. In the United States it was provided by the government-supported labs at the National Institutes of Health (NIH) and elsewhere. In Britain, first government grants, and then a large private medical research charity, the Wellcome Trust, paid for the research. The French and Japanese governments funded smaller laboratories, making the project truly international.

Second, the project – and indeed, modern science itself – would be impossible without the computer. The scientists had to analyse large amounts of information as they looked at each strand of DNA and tried to see where it began and ended. For humans, it would be overwhelming, but computers do this quickly. Many scientific projects now include people who only look after the computers and computer programmes, not the fruit flies or test tubes.

Third, modern science is big business, with a lot of money to be made as well as spent. The Human Genome Project became a race between the publicly funded groups and a private company established by the American entrepreneur Craig Venter (b. 1946). Venter, a gifted scientist, helped develop some of the equipment that could speed up DNA sequencing. He wanted to be the first to decode the human genome, patent his knowledge and charge scientists and pharmaceutical companies to use his information. The final result was a compromise. The whole human genome is freely available, but some of the ways that this information can be used can be patented, and the resulting drugs or diagnostic tests can be sold for profit. And, of course, people today pay to have their DNA sequenced, hoping that what they learn will help them maintain their health and avoid diseases that might affect them in the future.

Finally, the genome project is a telling example of the 'hype' surrounding today's important science. Scientists must compete for scarce funds, and sometimes exaggerate the significance of their research to get their grants. Journalists cover their stories, putting the most dramatic gloss they can on them, since ordinary science is not news. Each fresh announcement of a discovery or breakthrough raises the public's expectations that a cure or treatment is just around the corner. But mostly science takes longer for its lasting effects to be realised. New knowledge *is* gained every day, and new therapies *are* regularly introduced. But most science advances little by little, and media hype is rarely spot-on.

Yet it is a huge achievement to be able to read the human genome, because it can give us a much more precise understanding of health and disease. It will, in time, help us to develop new drugs against cancer, heart disease, diabetes, dementia and the other killers of modern times. We all stand to lead healthier lives as a result of this important work, involving scientists in many fields and many countries.

CHAPTER 39

The Big Bang

If a film of the history of the universe had been made, what would happen if you ran it backwards? At about five billion years ago our planet would disappear, for this is when it probably formed, from the debris of our solar system. Keep going back to the beginning and what happened then? The Big Bang: an explosion so powerful that its temperature and force are still being felt some 13.8 billion years later.

At least this is what scientists from the 1940s began to suggest with increasing confidence. The universe had begun from a point, an unimaginably hot, dense state, and then there was the big bang. Ever since this moment, it has been cooling and expanding, carrying the galaxies outwards from this original point. Ours is a dynamic and exciting universe, in which we are the tiniest of tiny specks. It is composed of the stars, planets and comets making up the visible galaxies; there is also much that's invisible – black holes and the much more abundant 'dark matter' and 'dark energy'.

So, did the Big Bang really happen, and can it explain the universe? Nobody was there of course, to begin filming. And what happened just before the Big Bang? These are questions that it is impossible to answer with any certainty, but they involve a lot of cutting-edge physics, as well as cosmology (the study of the universe). They have generated much debate over the past half-century or so. And it goes on right now.

Around 1800, the French Newtonian, Laplace, developed his nebular hypothesis (Chapter 18). He was mainly aiming to argue that the solar system had developed from a giant gas cloud. It convinced a lot of people that the earth had an ancient history, which would help explain its characteristics, such as its central heat, fossils and other geological features. Many nineteenth-century scientists passionately disputed the age of the earth and of our galaxy, the Milky Way. In the early decades of the twentieth century, two developments radically altered the questions.

The first was Einstein's General Theory of Relativity, with its important implications for time and space (Chapter 32). By insisting that these two things are intimately related, as 'space-time', Einstein added a new dimension to the universe. Einstein's mathematical work also implied that space was curved, so that Euclid's geometry didn't quite provide an adequate explanation over the vast distances of space. In Euclid's universe, parallel lines go on for ever, and never touch. But this assumes that space is flat. In a flat, Euclidian world, the sum of the angles of a triangle is always 180 degrees. But if you are measuring a triangle on a globe, with its curved surface, this doesn't work. And if space itself is curved, we need different forms of mathematics to deal with it.

Having accepted the essential truth of Einstein's brilliant work, the physicists and cosmologists had some new thinking to do. While the revolution he brought about was largely a theoretical one, the second major development in cosmology was not theoretical. It was based firmly on observations, especially those of the American astronomer Edwin Hubble (1889–1953). Hubble was celebrated in 1990 when a space shuttle carried into orbit round the earth a space telescope named after him. The Hubble Space

Telescope has recently revealed more than even he could have seen
with the telescope at the Mount Wilson Observatory in California,
where he worked. In the 1920s, Hubble saw further than any
astronomer had ever done. He showed that our galaxy (the Milky
Way) is not even the beginning of the end of the universe. It is one
of countless thousands of other galaxies, stretching even farther
than our telescopes can reach.

Cosmologists also remember Hubble for the special number, the
'constant', attached to his name. (You may remember Planck's
constant, which was a similar idea.) When light is moving away
from us, it shifts the spectrum of its waves to the red end of the
visible spectrum. This is called the 'red shift'. If it is moving towards
us, its waves shift towards the other end of the spectrum, the 'blue
shift'. This is an effect that astronomers can easily measure, and is
caused by the same thing that makes trains sound different when
they are coming towards you and going away from you. What
Hubble saw is that light from very distant stars has red shifts,
and the further away the star is, the larger the shift. This told
him that the stars are moving away from us, and the further away
they are, the faster they are moving. The universe is expanding,
and it appears to be doing so at an increasing rate. Hubble measured the distance from the stars and the extent of the red shift. His
measurements fell on a pretty straight line when he plotted them
on a graph. From this he calculated 'Hubble's constant', which he
published in a very important paper in 1929. This extraordinary
number gave cosmologists a method of calculating the age of the
universe.

Hubble's constant has been refined since then. New observations
have found stars even farther away, and we can now make more
accurate measurements of the red shift. Some of these stars are
millions of light years away. A light year is about six trillion earth
miles. It takes only eight minutes for a ray of sunlight to reach the
earth. If the ray of light then bounced back to the sun, it could
make over 32,000 return journeys in a year – another way of trying
to appreciate the vast distances involved. And vast amounts of
time. Some of what we see in the night sky is light that began its

journey a very long time ago from stars that have since become extinct. To get a really precise value for Hubble's constant, we need to know exactly how far away these very distant stars and galaxies are from us. But even with these difficulties, the constant's importance is that it can tell us approximately how long they have been travelling. This gives the age of the universe – beginning with its Big Bang.

The Big Bang was popularised in the 1940s by George Gamow (1904–68). Gamow was a colourful Russian-born physicist who went to America in the early 1930s. He had a wonderfully creative mind, contributing ideas to molecular biology as well as physics and relativity theory. With a colleague, he explored, at the micro-level, how the nucleus of an atom emits electrons (beta particles). On the grand scale, he looked at how nebulae – massive clouds of hot particles and cosmic dust – are formed. His theory of the Big Bang, worked out from 1948 with others, built on knowledge of the smallest constituents of atoms, combined with a model of what might have happened when the universe began.

First, the constituents: the particles and forces. In the late 1940s this bit of physics came to be called *quantum electrodynamics* or QED for short. One man who helped make sense of it was the American physicist Richard Feynman (1918–88). He is famous for the diagrams he drew (sometimes on restaurant napkins) to explain his theories and his mathematics, and for playing the bongo drums. He won the Nobel Prize in 1965, primarily for his work on QED, which provided the complicated mathematics to describe the even smaller particles and forces that we examine below.

After the end of the Second World War, particle physicists continued to accelerate atoms and then particles in increasingly more powerful particle accelerators. The accelerators can break up atoms into their sub-atomic particles, which is like reversing what might have happened a few instants after the Big Bang. Immediately after the Big Bang, as cooling began, the building-blocks of matter would have begun to form. From the particles would come the atoms and from the atoms the elements, and so on up to the planets and stars.

As Einstein's $E = mc^2$ tells us, at ever-higher speeds – almost the speed of light – in the accelerators, the mass is mostly converted into energy. The physicists found that these very fast particles do some fascinating things. The electron emerges unchanged from the accelerator. It is part of a family of force-particles – the *leptons*. The proton and neutron turn out to be composed of even smaller particles called *quarks*. There are several kinds. Each comes with a charge. Combined into threes, they make up a neutron or a proton.

There are four basic forces in the universe. Understanding how they relate to each other has been one of the great quests of the twentieth century. *Gravity* is the weakest, but acts at an infinite distance. It is still not entirely understood, even though we have been officially puzzling about it since Newton's apple. *Electromagnetism* is involved in many aspects of nature. It keeps the electrons in their orbits in the atom, and, as light, brings us daily news that the sun is still shining. Also in the atom are the *strong* and *weak nuclear forces*. These two bind the particles within the nucleus of the atom.

Leaving aside gravity, the other forces work by the exchange of special particles – force carriers – called *bosons*. These include the photon, Einstein's quantum of light, which is the boson for electro-magnetism. Yet, perhaps the most famous boson is the missing one: the Higgs Boson. Particle physicists have been looking it for since the 1960s. This boson is thought to create mass in other particles. Finding it would help explain how particles gained their mass in the immediate aftermath of the Big Bang. At the world's biggest particle accelerator, the Large Hadron Collider (LHC) near Geneva, Switzerland, scientists think they caught a glimpse of it on their instruments in 2012. The LHC was constructed between 1998 and 2008 by the European Organisation for Nuclear Research (CERN). CERN itself was established in 1954. It was a cooperative scientific enterprise among several European countries, a result of the high cost of physics research, and the need for many scientists, technicians and computer staff to perform and interpret these experiments at the extremes of matter and energy.

The Higgs Boson would be an extremely useful (but not the final) part of the puzzle known as the Standard Model, which accounts for everything except gravity. And a confirmed Standard Model would move close to a 'Theory of Everything', possibly via string theory, an approach to analysing all these forces and particles. String theory is based on the assumption that these fundamental forces of nature can be considered as if they were one-dimensional vibrating strings. It uses very complicated mathematics. This work is still science in the making.

A lot of this micro-level particle physics is difficult to associate with the ordinary world we live in. But scientists are finding more and more uses for it in nuclear energy, television, computers, quantum computing and medical screening equipment. Beyond these important uses in our daily lives, there is much to be learned too as the idea of the Big Bang has been fitted into what can be seen and not seen in the far reaches of space.

In the 1920s, the Russian physicist Alexander Friedman (1888–1925) was one of those who quickly assimilated Einstein's general theory of relativity into his own mathematical understanding of the universe. His Friedman Equations provided rules for an expanding universe. Friedman also wondered if it mattered that we looked out at the stars from earth. It's a special place for us, but did this give us a unique place for seeing the universe? He said no, it didn't matter. It's just where we happen to be. Things would not look different if we were on some other planet, light years away. This is Friedman's Cosmological Constant. It gives us another important idea: that matter is uniformly distributed throughout the universe. There are local variations, of course – the earth is much denser than the surrounding atmosphere. But smoothed out across all space, the principle appears to be true. Today, cosmologists still base much of their exploration on Friedman's models. They also have to deal with mysterious things such as black holes and dark matter.

Two fellows of the Royal Society discussed the idea of a 'dark star' in the eighteenth century. Describing its modern equivalent, the 'black hole', was the work of a modern mathematical genius,

Roger Penrose (b. 1931), and a brilliant theoretical physicist, Stephen Hawking (b. 1942). Until his retirement, Hawking had Isaac Newton's old job as Lucasian Professor of Mathematics at the University of Cambridge. Together they explained how black holes are easy to imagine, but of course impossible to see. This is because they are caused by areas in space where dying stars have gradually shrunk. As their remaining matter becomes more densely packed, the forces of gravity become so strong that the photons of light are trapped and cannot get out.

There are also super-massive black holes. In 2008 the Milky Way's very own super black hole – Sagittarius A* – was confirmed after a sixteen-year hunt with telescopes in Chile. Astronomers led by the German Reinhard Genzel (b. 1952) watched the patterns of the stars that orbit the black hole at the centre of the galaxy. They used measurements of infra-red light because there is so much stellar dust between the black hole and us, 27,000 light years away.

These super-massive black holes might play a part in the formation of galaxies and involve another part of space we cannot see directly: dark matter. Dark matter is thought to account for much more of the universe – 80 per cent of its matter – than the 4 per cent of the visible stars and planets together with gas and space dust. Dark matter was first considered in the 1930s, to explain why large bits of the universe did not behave exactly as predicted. Scientists had realised there was a mismatch between the mass of the visible parts and their gravitational effects: something was missing. In the 1970s, the astronomer Vera Rubin (b. 1928) charted how fast stars on the edge of galaxies were moving. They were travelling faster than they should have been. Traditionally it was thought that the further they were away from the centre of the galaxy, the slower they would orbit. Dark matter would provide the extra gravity needed to speed up the stars. So indirectly evidence of dark matter was provided and it has been generally accepted. But what dark matter *is* remains a mystery – something else to be found or disproved in the future.

Modern cosmology has emerged from Einstein's theories, from thousands upon thousands of observations, with computers to analyse the data, and from Gamow's idea of the Big Bang. Like any good theory in science, the Big Bang has changed since Gamow's time. In fact, for two decades after it was put forward in 1948, physicists hardly concerned themselves with the origins of the universe. The Big Bang had to contend with another model of the universe, called the 'steady state' one, most associated with the astronomer Fred Hoyle (1915–2001). Hoyle's model enjoyed some backing in the 1950s. It suggested an infinite universe, with the continuous creation of new matter. In this mode, the universe has no beginning and no end. There were so many difficulties with the steady-state idea that it had only a brief scientific life.

Physicists now have information about short-lived particles and forces gathered in particle accelerators. They have observations in the far reaches of space. They have been able to refine what we know about the Big Bang. There is still a lot of disagreement about details, and even about some of the fundamental principles, but this is not unusual in science. The Big Bang model can make sense of much that can now be measured, including the red shifts of distant stars, background cosmic radiation and the fundamental atomic forces. It can accommodate black holes and dark matter. What the model does not do, is say *why* the Big Bang happened. But, then, science deals with the how, not the why. As in all branches of science, some physicists and cosmologists have religious beliefs and others do not. That is how it should be. The best science is done in an atmosphere of tolerance.

CHAPTER 40

Science in Our Digital Age

The next time you switch on your computer, you probably won't 'compute'. You might look up something, email your friends, or check the latest football score. But computers were originally machines that could only compute – calculate – things faster or more accurately than our brains can.

We think of computers as cutting-edge technology, but the idea of the computer is very old. In the nineteenth century, a British mathematician, Charles Babbage (1792–1871), devised a calculating machine that could be 'programmed' to do tricks. For instance, he could set it up to count by single numbers to 1,000,000, and then when it got there, skip to 1,000,002. Anyone who had patiently watched it count to 1,000,000 would have been surprised by the missing number. Babbage's point was that his machine could do things that we wouldn't expect in the normal run of nature.

In the late 1800s, the American mathematician Herman Hollerith (1860–1929) invented an electric machine that used punched cards to analyse lots of data. If the cards were punched correctly and fed into the machine, it could 'read' them and process the information. The Hollerith Machine was very useful in analysing the information that people put down on their census forms, gathered to help the government understand more about the population. Very quickly, it could compute basic data such as how much people earned, how many people lived in each household, and their ages and sexes. The punch card remained the way most computers worked until the Second World War.

During that war, computers came into their own for military purposes. They could calculate how far shells would travel, and they served a more dramatic role in the top-secret attempts to decode enemy messages. The Germans, British and Americans all developed computers to aid wartime security. Here is a wonderful irony: the modern computer has opened up everyone's world, but it began as something that only a very few people, with the highest security clearance, had access to.

The British and Americans used computers to analyse German coded messages. The heart of the British effort to break the German codes was an old country house called Bletchley Park, in Buckinghamshire. The Germans used two code-making (cipher) machines, Enigma and Lorenz. Each day the codes were changed, which demanded great adaptability from the decoding machines. The British designed two code-breaking machines, the Bombe and the Colossus. The Colossus was well named, for these computers were enormous machines, filling entire rooms and consuming large amounts of electricity. The computers used a series of vacuum tubes to switch the electrical signals. These tubes generated a great deal of heat and were constantly failing. Wide aisles separated the rows of tubes so that the technicians could easily replace the burnt-out filaments. In those days, 'debugging' didn't mean running a software program, it meant reaching in and clearing out the bugs – moths or flies – that had flown into the hot glass tube and shorted out the system. The code-breakers shortened the duration of the war and undoubtedly helped the Allies to win it.

A remarkable mathematician worked at Bletchley Park: Alan Turing (1912–54). He was educated at my old college in Cambridge, King's College, where his brilliance was recognised as a student there in the early 1930s. He was publishing important ideas on computer mathematics, and his work at Bletchley Park was outstanding. After the war he continued to push his ideas. He had great insights into the relationship between the way computers work and the way our brains work; on 'artificial intelligence' (AI); and even on developing a machine that could play chess. Chess grandmasters still usually win against a computer, but the machines are getting better at making the best move. Turing developed an early electronic computer called ACE at the National Physical Laboratory in Teddington, London. It had much greater computing capacity. His life had a tragic end. He was gay at a time when homosexual activity was illegal in Britain. Arrested by the police, he underwent a treatment with sex hormones, to 'cure' his sexual orientation. He almost certainly committed suicide by eating an apple laced with the poison strychnine. His life and death are reminders that outstanding scientists can be anyone of any race, gender, religion and sexual preference.

The enormous machines built during the war were valuable, but they were limited by those overheating valves. Next came an invention that has changed the computer and much else: the transistor. Developed from late 1947 by John Bardeen (1908–91), Walter Brattain (1902–87) and William Shockley (1910–89), this device can amplify and switch electronic signals. Transistors were much smaller than vacuum tubes and generated much less heat. They have made all kinds of electrical appliances, such as transistor radios, much smaller and more efficient. The three men shared the Nobel Prize in Physics for their work, and Bardeen went on to win a second one for his research on 'semi-conductors', the material that makes transistors and modern circuits possible.

The military continued to develop computing during the Cold War of 1945 to 1991. The two great superpowers, the USA and the USSR, distrusted each other, despite having been allies during the Second World War. Computers were used to analyse the data each

A LITTLE HISTORY OF SCIENCE

country collected about the other's activities. But increasingly powerful number-crunching computers were a great help to scientists, too. Physicists made the greatest use of these new and improving machines during the 1960s. High-energy particle accelerators created so much data that it would have been impossible for an army of people with pencils and paper to make sense of it all.

More and more, computer scientists became members of a range of scientific teams, and research budgets included their salaries and equipment. So it made a lot of sense if one team could speak to another not just person to person, but computer to computer. After all, the telephone had been around for a century, and sending messages by telegraph wires was even older. Then, in the early 1960s, 'packet switching' was invented. Digital messages could be broken up into smaller units, and each unit would travel by the easiest route, and then be reassembled at its destination, the receiving computer screen. When you are talking on a landline, you're in 'real time', and no one else can call you. But you can send or receive a message on a computer – an email or a post on a website – and it will be available whenever someone wants to read it.

Packet switching was developed simultaneously in the USA and the UK. As a feature of national security, it allowed military or political leaders to communicate with each other, and would work even if some of the communication facilities had been destroyed. Packet switching made it easier to connect groups of computers: *networking* them. Again, the earliest non-military groups to network were the scientists. So much modern science benefits from collaboration. Academic communities were the main beneficiaries of the ever-smaller and ever-faster computers of the 1960s. They were extremely large, extremely slow and extremely expensive, compared with what we use today. But you will be relieved to know it was possible to play computer games even then, so the fun started early. The pace of change in computing accelerated in the 1970s. Computers – or microcomputers, as they were called – with a screen and keyboard became small enough to fit onto a desk. As the microprocessor chips they contained became more powerful, the personal computer revolution began. Much of the research was done in Silicon Valley in California in the USA.

Computers continued to change the way academic communities worked and communicated with each other. One of the largest collections of physicists in the world worked at the European Organisation for Nuclear Research (CERN), which houses the Large Hadron Collider, the world's fastest particle accelerator (Chapter 39). Computer specialists at CERN took networking and data analysis to new heights in the 1980s and 1990s. One expert was Tim Berners-Lee (b. 1955). Berners-Lee was always fascinated by computers. He grew up with them, as both his parents were early computer pioneers. Berners-Lee studied physics at Oxford and then went to work at CERN. In 1989, he asked for some research funds for 'Information Management'. His bosses at CERN gave him some help, but he persisted with his idea of making the increasing amounts of information available on the Internet easily accessible to anyone with a computer and a telephone line. Along with his colleague Robert Cailliau (b. 1947), he invented the World Wide Web. At first it was used just at CERN and one or two other physics laboratories. Then, in 1993 it went public. This coincided with the massive growth of personal computers not just at work but in the home. People who led the personal computer revolution, like Microsoft's Bill Gates (b. 1955) and Apple's Steve Jobs (1955–2011) are modern scientific heroes (and became very rich). So 1955 turned out to be a good year for computers: Berners-Lee, Gates and Jobs were all born then.

The speed of computer development from the 1970s matched the rate of invention of methods for sequencing the genome. It's no coincidence that the two events happened at the same time. Modern science is unthinkable without the modern computer. Many fundamental scientific problems, from designing new drugs to modelling climate change, depend on these machines. At home, we use them for doing homework, booking holiday tickets, playing computer games. Embedded computer systems fly our aeroplanes, assist medical imaging, and wash our clothes. Like modern science, modern life is computer-based.

We shouldn't be surprised at this. One of the things I have tried to show in this little book is that at any moment of history, the science has been a product of that particular moment. Hippocrates' moment was different from Galileo's, or Lavoisier's. They dressed, ate and thought like other people at the time. The people in this book thought more sharply than most of their contemporaries, and were able to communicate their ideas. That is why what they thought and wrote is worth remembering.

Yet the science of our day is more powerful than ever before. Computers are good for criminals and hackers as well as scientists and students. Science and technology can be abused as easily as they can be used for our common good. We need good scientists, but we also need good citizens who will ensure that our science will make the world a better place for us all to live in.

Index